GLOBAL WARMING

An International Emerging Issue

GLOBAL WARMING

An International Emerging Issue

Parikshit Ballabh

CYBER TECH PUBLICATIONS
4264/3, Ansari Road, Daryaganj, New Delhi-110002
Ph.: 011-23244078. Fax: 011-23280028
E-Mail: cyberpublicationsdelhi@yahoo.com

Global Warming : An International Emerging Issue

First Edition 2009

Published by G.S. Rawat
For Cyber Tech Publications
4264/3, Ansari Road, Daryaganj, New Delhi-110002
Ph.: 011-23244078. Fax: 011-23280028
E-Mail: cyberpublicationsdelhi@yahoo.com

Printed at
Asian Offset Printers

Preface

Global warming refers to climate change that causes an increase in the average temperature of the lower atmosphere. Global warming can have many different causes, but it is most commonly associated with human interference, specifically the release of excessive amounts of greenhouse gases.

Uncontrolled economic activity, rapid population growth and urbanization are primarily responsible for environmental damage. Since last few decades, there has been widespread concern about the dangers of environmental damage worldwide. The environment moved form being the concern of few scientists, administrators, and conservation groups to being the focus of a mass movement.

Greenhouse gases, such as carbon dioxide (CO_2), methane (CH_4), water vapor, and fluorinated gases, act like a greenhouse around the earth. This means that they let the heat from the Sun into the atmosphere, but do not allow the heat to escape back into space. The more greenhouse gases there are, the larger the percentage of heat that is trapped inside the earth's atmosphere. The earth could not exist in its present state (that is, with life) without the presence of some naturally occurring greenhouse gases, such as CO_2, CH_4, and water vapor. Without any greenhouse gases no heat would be trapped in atmosphere, so the earth would be extremely cold.

It is hoped that this work will be helpful in understanding the various contentious issues, which have erupted during various climate change negotiations in Indian context.

Editor

Contents

1

What is the Global Warming

Definition

Climate change is any substantial change in Earth's climate that lasts for an extended period of time. Global warming refers to climate change that causes an increase in the average temperature of the lower atmosphere. Global warming can have many different causes, but it is most commonly associated with human interference, specifically the release of excessive amounts of greenhouse gases. (EPA, 2006)

Greenhouse gases, such as carbon dioxide (CO_2), methane (CH_4), water vapor, and fluorinated gases, act like a greenhouse around the earth. This means that they let the heat from the Sun into the atmosphere, but do not allow the heat to escape back into space. The more greenhouse gases there are, the larger the percentage of heat that is trapped inside the earth's atmosphere. The earth could not exist in its present state (that is, with life) without the presence of some naturally occurring greenhouse gases, such as CO_2, CH_4, and water vapor. Without any greenhouse gases no heat would be trapped in atmosphere, so the earth would be extremely cold. (NASA, 2002)

Naturally occurring greenhouse gases (not fluorinated gases) are good in naturally occurring amounts; it's when people start contributing excessive amounts of them that greenhouse gases become a problem. With excessive greenhouse gas buildup, the earth's atmosphere warms to unnatural temperatures which causes, among other things, sea level to rise. Global warming also causes sea surface temperatures to rise, precipitation patterns to change, etc.. (NASA, 2002)

Global warming is the term used to describe a gradual increase in the average temperature of the Earth's atmosphere and its oceans, a change that is believed to be permanently changing the Earth's climate forever.

While many view the effects of global warming to be more substantial and more rapidly occurring than others do, the scientific consensus on climatic changes related to global warming is that the average temperature of the Earth has risen between 0.4 and 0.8 °C over the past 100 years. The increased volumes of carbon dioxide and other greenhouse gases released by the burning of fossil fuels, land clearing, agriculture, and other human activities, are believed to be the primary sources of the global warming that has occurred over the past 50 years.

Scientists from the Intergovernmental Panel on Climate carrying out global warming research have recently predicted that average global temperatures could increase between 1.4 and 5.8 °C by the year 2100. Changes resulting from global warming may include rising sea levels due to the melting of the polar ice caps, as well as an increase in occurrence and severity of storms and other severe weather events.

Global Warming is an International Issue

The warming of Earth's surface and oceans over the past century is very well documented, and climate research shows that most of the warming in the past half century results from manmade greenhouse gases.

Antarctica's Queen Maud Land

In recent years, global warming has been the subject of a great deal of political controversy. As scientific knowledge has grown, this debate is moving away from whether humans are causing warming and toward questions of how best to respond.

Signs that the earth is warming are recorded all over the globe. The easiest way to see increasing temperatures is through the thermometer records kept over the past century and a half. Around the world, the earth's average temperature has risen more than 1 degree Fahrenheit (0.8 degrees Celsius) over the last century, and about twice that in parts of the Arctic.

This doesn't mean that temperatures haven't fluctuated among regions of the globe or between seasons and times of day. But if you average out the temperature all over the world over the course of a year, you see that temperatures have been creeping upward.

Although we can't look at thermometers going back thousands of years, we do have some records that help us figure out what temperatures and concentrations were like in the distant past. For example, trees store information about the climate in the place where they live. Each year, trees grow thicker and form new rings. In warmer and wetter years, the rings are thicker. Old trees and wood can tell us about conditions hundreds or even several thousands of years ago.

Keys to the past are also buried under lakes and oceans. Pollen, creatures and particles fall to the bottom of oceans and lakes each year, forming sediments. Sediments preserve all these bits and pieces, which contain a wealth of information about what was in the air and water when they fell. Scientists reveal this record by inserting hollow tubes into the mud to collect sediment layers going back millions of years.

For a direct look at the atmosphere of the past, scientists drill cores through the earth's polar ice sheets. Tiny bubbles trapped in the gas are actually pieces of the earth's past atmosphere, frozen in time. That's how we know that the concentrations of greenhouse gases since the industrial revolution are higher than they've been for hundreds of thousands of years.

Computer models help scientists to understand the Earth's climate, or long-term weather patterns. Models also allow scientists to make predictions about the future climate. Basically, models simulate how the atmosphere and oceans absorb energy from the sun and transport it around the globe. Factors that affect the amount of the sun's energy reaching Earth's surface are what drive the climate in these models, as in real life. These include things like greenhouse gases, particles in the atmosphere (such as from volcanoes), and changes in energy coming from the sun itself.

2

Effects of Global Warming

Projected climate changes due to *global warming* have the potential to lead to future large-scale and possibly irreversible changes in our climate resulting in impacts at continental and global scales.

Examples of projected climate changes include:

* significant slowing of the ocean circulation that transports warm water to the North Atlantic,
* large reductions in the Greenland and West Antarctic Ice Sheets,
* accelerated global warming due to carbon cycle feedbacks in the terrestrial biosphere, and
* releases of terrestrial carbon from permafrost regions and methane from hydrates in coastal sediments.

The likelihood of many of these changes is uncertain. However, the probability of one or more of these changes occurring is likely to increase with the rate, magnitude, and duration of climate change.

The effects of global warming are not uniformly negative. Global warming will lead to climate change which will have positive benefits in some regions and negative effects in others. Scientists are unable to accurately predict when various effects of global warming will occur or what the magnitude of the effect will be.

For this reason, it is not possible to be certain whether the positive benefits will outweigh the negative impacts. What is known is that some significant negative impacts are projected and these drive most of the concern about global warming and motivates attempts to mitigate or adapt to the effects of global warming.

Most of the consequences of global warming would result from one of three physical changes: sea level rise, higher local temperatures, and changes in rainfall patterns (Figure 2.1). Sea level is generally expected to rise 50-200 cm in the next century (Dean et al. 1987); such a rise would inundate 7,000 square miles of dry land in the United States (an area the size of Massachusetts) and a similar amount of coastal wetlands; erode recreational beaches 100-200 meters, exacerbate coastal flooding; and increase the salinity of aquifers and estuaries (Titus 1989).

Effects on Weather

Increasing temperature is likely to lead to increasing

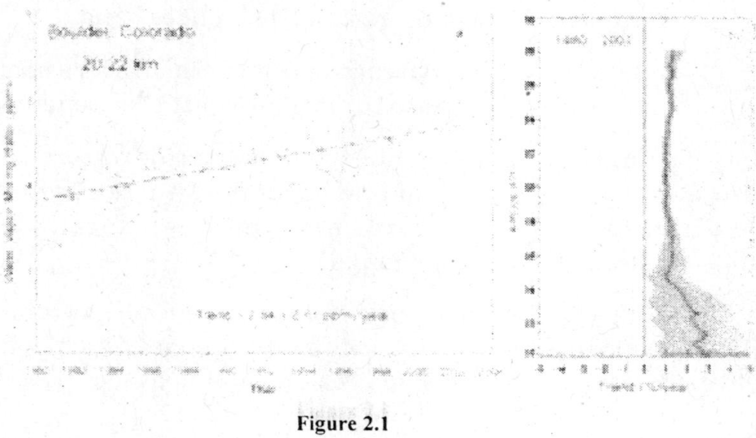

Figure 2.1

precipitation but the effects on storms are less clear. Extratropical storms partly depend on the temperature *gradient*, which is predicted to weaken in the northern hemisphere as the polar region warms more than the rest of the hemisphere.

More extreme weather

The Intergovernmental Panel on Climate Change *(IPCC) Third Assessment Report* "Climate Change 2001" stated "there is no compelling evidence to indicate that the characteristics of tropical and extratropical storms have changed." There is, however, limited evidence from a relatively short time period that storm strength is increasing, such as the Emanuel (2005) "power dissipation index" of hurricane intensity. Worldwide, the proportion of *hurricanes* reaching *categories 4 or 5* – with wind speeds above 56 metres per second – has risen from 20% in the 1970s to 35% in the 1990s. Precipitation hitting the US from hurricanes increased by 7% over the twentieth century. See also *Time Magazine's "Global Warming: The Culprit?" and.*

The World Meteorological Organization has suggested a possible link between global warming and increasing extreme weather events, as have Hoyos *et al.* (2006), writing, "the increasing ... number of category 4 and 5 hurricanes ... is directly linked to" increasing temperatures. Hurricane modeling has produced similar results, e.g., "hurricanes, simulated under warmer, high-CO_2 conditions, are more intense ... than under present-day conditions.... greenhouse gas–induced warming may lead to ... increasing ... occurrence of highly destructive category-5 storms." A paper by 14 scientists notes that "research shows very little evidence to support the claim that the rising costs associated with weather ... are associated with changes in [their] frequency or intensity." The IPCC TAR, in 2001, made no strong statements on the issue.

Destabilization of local climates

Figure 2.2: The first recorded South Atlantic hurricane, "Catarina", which hit Brazil in 2004

A substantially higher risk of extreme weather does not necessarily mean a noticeably greater risk of slightly-above-average weather. However, the evidence is clear that severe weather and moderate rainfall are also increasing.

Stephen Mwakifwamba, national co-ordinator of the Centre for Energy, Environment, Science and Technology - which prepared the Tanzanian government's climate change report to the UN - says that change is happening in *Tanzania* right now. "In the past, we had a drought about every 10 years", he says. "Now we just don't know when they will come. They are more frequent, but then so are floods. The climate is far less predictable. We might have floods in May or droughts every three years. Upland areas, which were never affected by mosquitoes, now are. Water levels are decreasing every day. The rains come at the wrong time for farmers and it is leading to many problems".

Greg Holland, director of the Mesoscale and Microscale Meteorology Division at the National Center for Atmospheric Research in Boulder, Colorado, said on April 24, 2006, "The hurricanes we are seeing are indeed a direct result of climate change," and that the wind and warmer water conditions that fuel storms that form in the Caribbean are, "increasingly due to greenhouse gases. There seems to be no other conclusion you can logically draw." Holland said, "The large bulk of the scientific community say what we are seeing now is linked directly to greenhouse gases."

Increased evaporation

Increasing water vapor at Boulder, Colorado.

As the climate grows warmer, evaporation will increase. This may cause heavier rainfall and more erosion, and in more vulnerable tropical areas (especially in Africa), desertification due to deforestation. Many scientists think that it could result in more extreme weather as global warming progresses. The IPCC Third Annual Report says: "...global average water vapour concentration and precipitation are projected to increase during the 21st century. By the second half of the 21st century, it is likely that precipitation will have increased over northern mid- to high latitudes and Antarctica in winter. At low latitudes there are both regional increases and decreases over land areas. Larger year to year variations in precipitation are very likely over most areas where an increase in mean precipitation is projected".

Cost of more extreme weather

The economic impact of extreme weather is rising

rapidly both because of increases in the frequency and intensity of extreme weather and because of changes in human behavior. An example of how human behavior has increased exposure to extreme weather is the movement towards greater development along vulnerable seacoasts. The economic impact of hurricanes has increased because there is more development along seacoasts vulnerable to hurricanes. Similarly, the economic impact of floods has increased because there is more development in floodplains.

Choi and Fisher, writing in *Climate Change,* vol. 58 (2003) pp. 149, predict that each 1% increase in annual precipitation would enlarge the cost of catastrophic storms by 2.8%.

The Association of British Insurers has stated that limiting carbon emissions would avoid 80% of the projected additional annual cost of tropical cyclones by the 2080s. The cost is also increasing partly because of building in exposed areas such as coasts and floodplains. The ABI claims that reduction of the vulnerability to some inevitable impacts of climate change, for example through more resilient buildings and improved ûood defences, could also result in considerable cost-savings in the longterm.

In the northern hemisphere, the southern part of the *Arctic* region (home to 4,000,000 people) has experienced a temperature rise 1° to 3° Celsius over the last 50 years. *Canada, Alaska* and *Russia* are experiencing initial melting of *permafrost*. This may disrupt ecosystems and by increasing bacterial activity in the soil lead to these areas becoming carbon sources instead of *carbon sinks*. A study (published in *Science*) of changes to eastern *Siberia's permafrost* suggests that it is gradually disappearing in the southern regions, leading to the loss of nearly 11% of Siberia's nearly 11,000 lakes since 1971. At the same time, western Siberia is at the initial stage where melting permafrost is creating new lakes, which will eventually start disappearing as in the east. Western Siberia is the world's largest *peat bog,* and the melting of its permafrost is likely to lead to the release, over decades, of large quantities of methane—creating an additional source of greenhouse gas emissions.

Hurricanes were thought to be an entirely north Atlantic phenomenon. In April 2004, the first Atlantic hurricane to form south of the Equator hit *Brazil* with 40 m/s (144 km/h) winds; monitoring systems may have to be extended 1,600 km (1000 miles) further south.

Oceans Sea level rise

Sea level has been rising 0.2 cm/year, based on measurements of sea level rise from 23 long tide gauge records in geologically stable environments

With increasing average global temperature, the *water* in the oceans expands in volume, and additional water enters them which had previously been locked up on land in glaciers and the *polar ice caps.* An increase of 1.5 to 4.5 °C is estimated to lead to an increase of 15 to 95 cm (IPCC 2001).

The sea level has risen more than 120 *metres* since the peak of the last *ice age* about 18,000 years ago. The bulk of that occurred before 6000 years ago. From 3000 years ago to the start of the 19th century, sea level was almost constant, rising at 0.1 to 0.2 mm/yr; since 1900, the level has risen at 1–2 mm/yr; since 1992, satellite altimetry from TOPEX/Poseidon indicates a rate of about 3 mm/yr.

Temperature rise

The temperature of the Antarctic *Southern Ocean* rose by 0.17 °C (0.31 °F) between the 1950s and the 1980s, nearly twice the rate for the world's oceans as a whole.

As well as effects on ecosystems (eg by melting sea ice, affecting algae that grow on its underside), warming could reduce the ocean's ability to absorb CO_2.

More important for the *United States* may be the temperature rise in the *Gulf of Mexico*. As hurricanes cross the warm *Loop Current* coming up from *South America*, they can gain great strength in under a day (as did *Hurricane Katrina* and *Hurricane* Rita in 2005), with water above 85 degrees F seemingly promoting Category 5 storms. Hurricane season ends in November as the waters cool.

Acidification

The world's oceans soak up much of the carbon dioxide produced by living organisms, either as dissolved gas, or in the skeletons of tiny marine creatures that fall to the bottom to become chalk or limestone. Oceans currently absorb about one metric tonne of CO_2 per person per year. It is estimated that the oceans have absorbed around half of all CO_2 generated by human activities since 1800 (120,000,000,000 tonnes or 120 *petagrams* of carbon).

But in water, carbon dioxide becomes a weak carbonic acid, and the increase in the greenhouse gas since the industrial revolution has already lowered the average pH (the laboratory measure of acidity) of seawater by 0.1 units on the 14-point scale, to 8.2. Predicted emissions could lower it by a further 0.5 by 2100, to a level not seen for millions of years.

There are concerns that increasing acidification could have a particularly detrimental effect on corals (16% of the world's coral reefs have died from bleaching since 1998 and other marine organisms with calcium carbonate shells Increased acidity may also directly affect the growth and reproduction of fish as well as the plankton on which they rely on for food.

Shutdown of thermohaline circulation

There is some speculation that global warming could, via a shutdown or slowdown of the thermohaline circulation, trigger localised cooling in the North Atlantic and lead to cooling, or lesser warming, in that region. This would affect in particular areas like Scandinavia and Britain that are warmed by the North Atlantic drift. The chances of this occurring are unclear; there is some evidence for the stability of the Gulf Stream and possible weakening of the North Atlantic drift. There is, however, no evidence for cooling in northern Europe or nearby seas; quite the reverse.

Ecosystems

Rising temperatures are beginning to have a noticeable impact on ecosystems. Secondary evidence of global warming — lessened snow cover, rising sea levels, weather changes — provides examples of consequences of global warming that may influence not only human activities but also the ecosystems. Increasing global temperature means that ecosystems may change; some species may be forced out of their habitats (possibly to extinction) because of changing conditions, while others may flourish.

Few of the terrestrial ecoregions on Earth could expect to be unaffected. Many of the species at risk are arctic fauna such as polar bears, emperor penguins, many salt wetland flora and fauna species, and any species that inhabit the low land areas near the sea. Species that rely on cold weather conditions such as gyrfalcons, and snowy owls that prey on lemmings that use the cold winter to their advantage will be hit hard.

Butterflies have shifted their ranges northward by 200 km in Europe and North America. Plants lag behind, and larger animals' migration is slowed down by cities and highways. In Britain, spring butterflies are appearing an average of 6 days earlier than two decades ago. In the Arctic, the waters of Hudson Bay are ice-free for three weeks longer than they were thirty years ago, affecting *polar bears,* which do not hunt on land.

Two 2002 studies in *Nature* (vol 421) surveyed the scientific literature to find recent changes in range or seasonal behaviour by plant and animal species. Of species showing recent change, 4 out of 5 shifted their ranges towards the poles or higher altitudes, creating "*refugee species*". Frogs were breeding, flowers blossoming and birds migrating an average 2.3 days earlier each decade; butterflies, birds and plants moving towards the poles by 6.1 km per decade. A 2005 study concludes human activity is the cause of the temperature rise and resultant changing species behaviour, and links these effects with the predictions of *climate models* to provide validation for them. Grass has become established in Antarctica for the first time.

Forests in some regions potentially face an increased risk of forest fires. The 10-year average of boreal forest burned in North America, after several decades of around 10,000 km^2 (2.5 million acres), has increased steadily since 1970 to more than 28,000 km^2 (7 million acres) annually. This change may be due in part to changes in forest management practices.

Ecological productivity

Increasing average temperature and carbon dioxide may have the effect, up to a point, of improving ecosystems' productivity. Atmospheric carbon dioxide is rare in comparison to oxygen (less than 1% of air compared to 21% of air). This carbon dioxide starvation becomes apparent in *photorespiration,* where there is so little carbon dioxide, that oxygen can enter a plant's chloroplasts and takes the place where carbon dioxide normally would be in the Calvin

Cycle. This causes the sugars being made to be destroyed, badly suppressing growth. Satellite data shows that the productivity of the northern hemisphere has increased since 1982 (although attribution of this increase to a specific cause is difficult).

IPCC models predict that higher CO_2 concentrations would only spur growth of flora up to a point, because in many regions the limiting factors are water or nutrients, not temperature or CO_2; after that, greenhouse effects and warming would continue but there would be no compensatory increase in growth.

Research done by the *Swiss Canopy Crane Project* suggests that slow-growing trees only are stimulated in growth for a short period under higher CO_2 levels, while faster growing plants like liana benefit in the long term. In general, but especially in *rain forests,* this means that liana become the prevalent species; and because they decompose much faster than trees their carbon content is more quickly returned to the atmosphere. Slow growing trees incorporate atmospheric carbon for decades.

Glacier Retreat

A map of the change in thickness of mountain glaciers since 1970. Thinning in orange and red, thickening in blue.

Lewis Glacier, North Cascades, WA USA is one of five glaciers in the area that melted away

In historic times, glaciers grew during the *Little Ice Age*, a cool period from about 1550 to 1850. Subsequently, until about 1940, glaciers around the world retreated as climate warmed. *Glacier retreat* declined and reversed, in many cases, from 1950 to 1980 as a slight global cooling occurred. Since 1980, glacier retreat has become increasingly rapid and ubiquitous, so much so that it has threatened the existence of many of the glaciers of the world. This process has increased markedly since 1995.

The total surface area of *glaciers* worldwide has decreased by 50% since the end of the 19th century. Currently glacier retreat rates and mass balance losses have been increasing in the Andes, Alps, Himalaya's, Rocky Mountains and North Cascades. As of March 2005, the snow cap that has covered the top of *Mount Kilimanjaro* for the past 11,000 years since the last *ice age* has almost disappeared.

The loss of glaciers not only directly causes landslides, flash floods and *glacial lake* overflow, but also increases annual variation in water flows in rivers. Glacier runoff declines in the summer as glaciers decrease in size, this decline is already observable in several regions. Glaciers retain water on mountains in high precipitation years, since the snow cover accumulating on glaciers protects the ice from melting. In warmer and drier years, glaciers offset the lower precipitation amounts with a higher meltwater input.

Of particular importance are the *Hindu Kush* and *Himalayan* glacial melts. The melt of these glaciers is a large and reliable source of water for *China, India,* and much of *Asia*, and these waters form the principal dry-season water source many of the regions major rivers, including the *Ganges*. Increased melting would cause greater flow for several decades, after which " some areas of the most populated regions on Earth are likely to 'run out of water'.

The recession of mountain glaciers, notably in Western North America, Franz-Josef Land, Asia, the Alps, Indonesia and Africa, and tropical and sub-tropical regions of South America, has been used to provide qualitative support to the rise in global temperatures since the late 19th century. Many glaciers are being lost to melting further raising concerns about future local water resources in these glacierized areas. The Lewis Glacier, North Cascades pictured at right after melting away in 1990 is one of the 47 North Cascade glaciers observed and all are retreating.

Despite their proximity and importance to human populations, the mountain and valley glaciers of temperate latitudes amount to a small fraction of glacial ice on the earth. About 99% is in the great ice sheets of polar and subpolar Antarctica and Greenland. These continuous continental-scale ice sheets, 3 km (1.8 miles) or more in thickness, cap the polar and subpolar land masses. Like rivers flowing from an enormous lake, numerous outlet glaciers transport ice from the margins of the ice sheet to the ocean.

Glacier retreat has been observed in these outlet glaciers, resulting in an increase of the ice flow rate. In Greenland the period since the year 2000 has brought retreat to several very large glaciers that had long been stable. Three glaciers that have been researched, Helheim, Jakobshavns and Kangerdlugssuaq Glaciers, jointly drain more than 16% of the Greenland Ice Sheet. Satellite images and aerial photographs from the 1950s and 1970s show that the front of the glacier had remained in the same place for decades. But in 2001 it began retreating rapidly, retreating 7.2 km (4.5 miles) between 2001 and 2005. It has also accelerated from 20 m (65 ft)/day to 32 m (104 ft)/day. *Jakobshavn Isbræ* in west Greenland is generally considered the fastest moving glacier in the world. It had been moving continuously at speeds of over 24 m (78 ft)/ day with a stable terminus since at least 1950. In 2002, the 12 km (7.5 mile) long floating terminus entered a phase of rapid retreat. The ice front started to break up and the floating terminus disintegrated accelerating to a retreat rate of over 30 m (98 ft)/day. The acceleration rate of retreat of Kangerdlugssuaq Glacier is even larger. Portions of the main trunk that were flowing at 15 m (49 ft)/day in 1988-2001 were flowing at 40 m (131 ft)/day in summer 2005. The front of the glacier has also retreated and has rapidly thinned by more than 100 m (328 ft).

Glacier retreat and acceleration is also apparent on two important outlet glaciers of the West Antarctic Ice Sheet. Pine Island Glacier, which flows into the Amundsen Sea thinned 3.5 ± 0.9 m (11.5 ± 3 ft) per year and retreated five kilometers (3.1 miles) in 3.8 years. The terminus of the glacier is a floating ice shelf and the point at which it is afloat is retreating 1.2 km/year. This glacier drains a substantial portion of the West Antarctic Ice Sheet and has been referred to as the weak underbelly of this ice sheet. This same pattern of thinning is evident on the neighboring Thwaites Glacier.

* Further global warming (positive feedback)
* Some effects of global warming themselves contribute directly to further global warming.
* Methane release from melting permafrost peat bogs
* Wikinews has news related to:

Scientists warn thawing Siberia may trigger global meltdown

Climate scientists reported in August 2005 that a one million square kilometer region of permafrost peat bogs in western Siberia is starting to melt for the first time since it was formed 11,000 years ago at the end of the last ice age. This will release methane, an extremely effective greenhouse gas, possibly as much as 70,000 million tonnes, over the next few decades. An earlier report in May 2005 reported similar melting in eastern Siberia.

This positive feedback was not known about in 2001 when the IPCC issued its last major report on climate change. The discovery of permafrost peat bogs melting in 2005 implies that warming is likely to happen faster than was predicted in 2001.

Methane release from hydrates

Methane clathrate, also called methane hydrate, is a form of water ice that contains a large amount of methane within its crystal structure. Extremely large deposits of methane clathrate have been found under sediments on the ocean floors of Earth. The sudden release of large amounts of natural gas from methane clathrate deposits has been hypothesized as a cause of past and possibly future climate changes. The release of this trapped methane is a potential major outcome of a rise in temperature; it is thought that this might increase the global temperature by an additonal 5° in itself, as methane is much more powerful as a greenhouse gas than carbon dioxide. The theory also predicts this will greatly affect availiable oxygen content of the atmophere. This theory has been proposed to explain the first mass extinction event on earth known as the Permian-Triassic extinction event.

Carbon cycle feedbacks

There have been predictions, and some evidence, that global warming might cause loss of carbon from terrestrial ecosystems, leading to an increase of atmospheric CO_2 levels. Several climate models indicate that global warming through the 21st century could be accelerated by the response of the terrestrial carbon cycle to such warming. All 11 models in the C4MIP study found that a larger fraction of anthropogenic CO2 will stay airborne if climate change is accounted for. By the end of the twenty-first century, this additional CO_2 varied between 20 and 200 ppm for the two extreme models, the majority of the models lying between 50 and 100 ppm. The higher CO2 levels led to an additional climate warming ranging between 0.1° and 1.5°C.

However, there was still a large uncertainty on the magnitude of these sensitivities. Eight models attributed most of the changes to the land, while three attributed it to the ocean. The strongest feedbacks in these cases are due to increased respiration of carbon from soils throughout the high latitude boreal forests of the Northern Hemisphere. One model in particular (HadCM3) indicates a secondary carbon cycle feedback due to the loss of much of the *Amazon rainforest* in response to significantly reduced precipitation over tropical South America. While models disagree on the strength of any terrestrial carbon cycle feedback, they each suggest any such feedback would accelerate global warming.

Observations show that soils in England have been losing carbon at the rate of four million tonnes a year for the past 25 years according to a paper in Nature by Bellamy et al. in September 2005, who note that these results are unlikely to be explained by land use changes. Results such as this rely on a dense sampling network and thus are not available on a global scale. Extrapolating to all of the United Kingdom, they estimate annual losses of 13 million tons per year. This is as much as the annual reductions in carbon dioxide emissions achieved by the UK under the Kyoto Treaty (12.7 million tons of carbon per year).

Forest Fires

Rising Global temperature might cause forest fires to occur on larger scale, and more regularly. This releases more stored carbon into the atmosphere than the carbon cycle can naturally re-absorb, as well as reducing the overall forest area on the planet, creating a positive feedback loop. Part of that feedback loop is more rapid growth of replacement forests and a northward migration of forests as northern latitudes become more suitable climates for sustaining forests. There is a question of whether the burning of renewable fuels such as forests should be counted as contributing to global warming.

- (Climate Change and Fire)
- (Climate Roulette: Loss of Carbon Sinks & Positive Feedbacks)
- (EPA: Global Warming: Impacts: Forests)
- (Feedback Cycles linking forests, climate and landuse activities)

Economic

In commenting overall economic effect of global warming in *Copenhagen Consensus*, Professor Robert O. Mendelsohn of Yale School of Forestry and Environmental Studies, stated that

"A series of studies on the impacts of climate change have systematically shown that the older literature overestimated climate damages by failing to allow for adaptation and for climate benefits (see Fankhauser et al 1997; Mendelsohn and Newmann 1999; Tol 1999; Mendelsohn et al 2000; Mendelsohn 2001;Maddison 2001; Tol 2002; Söhngen et al 2002; Pearce 2003; Mendelsohn and Williams 2004). These new studies imply that impacts depend heavily upon initial temperatures (latitude). Countries in the polar region are likely to receive large benefits from warming, countries in the mid-latitudes will at first benefit and only begin to be harmed if temperatures rise above 2.5C (Mendelsohn et al 2000). Only countries in the tropical and subtropical regions are likely to be harmed immediately by warming and be subject to the magnitudes of impacts first thought likely (Mendelsohn et al 2000). Summing these regional impacts across the globe implies that warming benefits and damages will likely offset each other until warming passes 2.5C and even then it will be far smaller on net than originally thought (Mendelsohn and Williams

2004)."

Decline of agriculture

For some time it was hoped that a positive effect of global warming would be increased agricultural yields, because of the role of carbon dioxide in photosynthesis, especially in preventing photorespiration, which is responsible for significant destruction of several crops. In Iceland, rising temperatures have made possible the widespread sowing of barley, which was untenable twenty years ago. Some of the warming is due to a local (possibly temporary) effect via ocean currents from the Caribbean, which has also affected fish stocks.

While local benefits may be felt in some regions (such as Siberia), recent evidence is that global yields will be negatively affected. "Rising atmospheric temperatures, longer droughts and side-effects of both, such as higher levels of ground-level ozone gas, are likely to bring about a substantial reduction in crop yields in the coming decades, large-scale experiments have shown" (The Independent, April 27, 2005, "Climate change poses threat to food supply, scientists say" - report on this event).

Moreover, the region likely to be worst affected is Africa, both because its geography makes it particularly vulnerable, and because seventy per cent of the population rely on rain-fed agriculture for their livelihoods. Tanzania's official report on climate change suggests that the areas that usually get two rainfalls in the year will probably get more, and those that get only one rainy season will get far less. The net result is expected to be that 33% less maize—the country's staple crop—will be grown.

An industry very directly affected by the risks is the insurance industry; the number of major natural disasters has trebled since the 1960s, and insured losses increased fifteen-fold in real terms (adjusted for inflation). According to one study, 35–40% of the worst catastrophes have been climate change related (ERM, 2002). Over the past three decades, the proportion of the global population affected by weather-related disasters has doubled in linear trend, rising from roughly 2% in 1975 to 4% in 2001 (ERM, 2002).

A *June 2004 report* by the Association of British Insurers declared "Climate change is not a remote issue for future generations to deal with. It is, in various forms, here already, impacting on insurers' businesses now". It noted that weather risks for households and property were already increasing by 2-4 % per year due to changing weather, and that claims for storm and flood damages in the UK had doubled to over £6 billion over the period 1998–2003, compared to the previous five years. The results are rising insurance premiums, and the risk that in some areas *flood risk insurance* will become unaffordable for some.

In the United States, insurance losses have also greatly increased, but according to one study those increases are attributed to increased population and property values in vulnerable coastal areas.

Transport

Roads, airport runways, railway lines and pipelines, (including oil pipelines, sewers, water mains etc) may require increased maintenance and renewal as they become subject to greater temperature variation, and, in areas with permafrost, subject to subsidence.

Flood defense

For historical reasons to do with trade, many of the world's largest and most prosperous cities are on the coast, and the cost of building better *coastal defenses* (due to the rising sea level) is likely to be considerable. Some countries will be more affected than others - low-lying countries such as *Bangladesh and the Netherlands* would be worst hit by any sea level rise, in terms of floods or the cost of preventing them.

In developing countries, the poorest often live on flood plains, because it is the only available space, or fertile agricultural land. These settlements often lack infrastructure such as dykes and early warning systems. Poorer communities also tend to lack the insurance, savings or access to credit needed to recover from disasters.

Migration

Some Pacific Ocean island nations, such as *Tuvalu,* are concerned about the possibility of an eventual evacuation, as flood defense may become economically inviable for them. Tuvalu already has an ad hoc agreement with *New Zealand* to allow phased relocation.

In the 1990s a variety of estimates placed the number of *environmental refugees* at around 25 million. (Environmental refugees are not included in the official definition of *refugees*, which only includes migrants fleeing persecution.) The Intergovernmental Panel on Climate Change (IPCC), which advises the world's governments under the auspices of the UN, estimated that 150 million environmental refugees will exist in the year 2050, due mainly to the effects of coastal flooding, shoreline erosion and agricultural disruption. (150 million means 1.5 percent of 2050's predicted 10 billion world population.)

Northwest Passage

Melting Arctic ice may open the Northwest Passage in summer, which would cut 5,000 nautical miles (9,000 km) from shipping routes between Europe and Asia. This would be of particular relevance for supertankers which are too big to fit through the Panama Canal and currently have to go around the tip of South America. According the

Canadian Ice Service, the amount of ice in Canada's eastern Arctic Archipelago decreased by 15 percent between 1969 and 2004.

While the reduction of summer ice in the Arctic may be a boon to shipping, this same phenomenon threatens the Arctic ecosystem, most notably polar bears which depend on ice floes. Subsistence hunters such as the Inuit peoples will find their livelihoods and cultures increasingly threatened as the ecosystem changes due to global warming.

Development

The combined effects of global warming may impact particularly harshly on people and countries without the resources to mitigate those effects. This may slow economic development and poverty reduction, and make it harder to achieve the Millennium Development Goals.

In October 2004 the *Working Group on Climate Change and Development*, a coalition of development and environment NGOs, issued a report Up in Smoke on the effects of climate change on development. This report, and the July 2005 report Africa - Up in Smoke? predicted increased hunger and disease due to decreased rainfall and severe weather events, particularly in Africa. These are likely to have severe impacts on development for those affected.

Environmental

Secondary evidence of global warming — reduced snow cover, rising sea levels, weather changes — provides examples of consequences of global warming that may influence not only human activities but also ecosystems. Increasing global temperature means that ecosystems may change; some species may be forced out of their habitats (possibly to extinction) because of changing conditions, while others may flourish. Few of the terrestrial ecoregions on Earth could expect to be unaffected.

Increasing carbon dioxide may (up to a point) increase ecosystems' productivity; but the interaction with other aspects of climate change, means the environmental impact of this is unclear. An increase in the total amount of biomass produced is not necessarily all good, since biodiversity can still decrease even though a smaller number of species are flourishing.

Water scarcity

Eustatic sea level rises threaten to contaminate groundwater, affecting drinking water and agriculture in coastal zones. Increased evaporation will reduce the effectiveness of reservoirs. Increased extreme weather means more water falls on hardened ground unable to absorb it - leading to flash floods instead of a replenishment of soil moisture or groundwater levels. In some areas, shrinking glaciers threaten the water supply.

Higher temperatures will also increase the demand for water for cooling purposes.

In the Sahel, there has been on average a 25 per cent decrease in annual rainfall over the past 30 years.

Health

Direct effects of temperature rise

Rising temperatures have two opposing direct effects on mortality: higher temperatures in winter reduce deaths from cold; higher temperatures in summer increase heat-related deaths. The distribution of these changes obviously differs. Palutikof et al calculate that in England and Wales for a 1 °C temperature rise the reduced deaths from cold outweigh the increased deaths from heat, resulting in a reduction in annual average mortality of 7000.

The European heat wave of 2003 killed 22,000–35,000 people, based on normal mortality rates (Schär and Jendritzky,

2004). It can be said with 90% confidence that past human influence on climate was responsible for at least half the risk of the 2003 European summer heat-wave (Stott et al 2004).

If average temperatures increase by 1 degree Celsius, there will be an estimated 24,000 additional murders in the U.S. each year (as the additional heat stress leads to more frequent rage). (New Scientist 11/5/02, review of *Body Heat* by Mark Blumberg.)

Spread of disease

Global warming is expected to extend the favourable zones for *vectors* conveying infectious disease such as malaria. In poorer countries, this may simply lead to higher incidence of such diseases. In richer countries, where such diseases have been eliminated or kept in check by vaccination, draining swamps and using pesticides, the consequences may be felt more in economic than health terms, if greater spending on preventative measures is required.

Impacts of glacier retreat

The continued retreat of glaciers will have a number of different impacts. In areas that are heavily dependent on water runoff from glaciers that melt during the warmer summer months, a continuation of the current retreat will eventually deplete the glacial ice and substantially reduce or eliminate runoff. A reduction in runoff will affect the ability to irrigate crops and will reduce summer stream flows necessary to keep dams and reservoirs replenished. This situation is particularly acute for irrigation in South America, where numerous artificial lakes are filled almost exclusively by glacial melt. Central Asian countries have also been historically dependent on the seasonal glacier melt water for irrigation and drinking supplies. In Norway, the Alps, and the Pacific Northwest of North America, glacier runoff is important for hydropower.

Many species of freshwater and saltwater plants and animals are dependent on glacier-fed waters to ensure a cold water habitat that they have adapted to. Some species of freshwater fish need cold water to survive and to reproduce, and this is especially true with Salmon and Cutthroat trout. Reduced glacier runoff can lead to insufficient stream flow to allow these species to thrive. Ocean krill, a cornerstone species, prefer cold water and are the primary food source for aquatic mammals such as the Sperm whale. Alterations to the *ocean currents*, due to increased freshwater inputs from glacier melt, and the potential alterations to thermohaline circulation of the worlds oceans, may impact existing fisheries upon which humans depend as well.

The potential for major *sea level rise* is mostly dependent on a significant melting of the polar ice caps of Greenland and Antarctica, as this is where the vast majority of glacial ice is located. The *British Antarctic Survey* has determined from climate modeling that for at least the next 50 years, snowfall on the continent of Antarctica should continue to exceed glacial losses from global warming. The amount of glacial loss on the continent of Antarctica is not increasing significantly, and it is not known if the continent will experience a warming or a cooling trend, although the *Antarctic Peninsula* has warmed in recent years, causing glacier retreat in that region. If all the ice on the polar ice caps were to melt away, the oceans of the world would rise an estimated 70 m (229 ft). However, with little major melt expected in Antarctica, sea level rise of not more than 0.5 m (1.6 ft) is expected through the 21st century, with an average annual rise of 0.0004 m (0.0013 ft) per year. *Thermal expansion* of the world's oceans will contribute, independent of glacial melt, enough to double those figures.

3

Reasons and Preventions of Global Warming

Global Warming Skeptics - Skeptics of global warming think that global warming is not an ecological trouble.

Global Warming Facts - 8 Facts about Global Warming

Causes of Global Warming - The Green house gases are the main culprits of the global warming. The green house gases like carbon dioxide, methane, and nitrous oxide are playing hazards in the present times.

Green House Gasses are the ingredients of the atmosphere that add to the greenhouse effect.

Al Gore Global Warming Initiative - Gore has written a book that archives his advice that Earth is dashing toward an immensely warmer future.

The average facade temperature of the globe has augmented more than 1 degree Fahrenheit since 1900 and the speed of warming has been almost three folds the century long average since 1970. This increase in earth's average temperature is called Global warming. More or less all specialists studying the climate record of the earth have the same opinion now that human actions, mainly the discharge of green house gases from smokestacks, vehicles, and burning forests, are perhaps the leading power driving the fashion.

The gases append to the planet's normal greenhouse effect, permitting sunlight in, but stopping some of the ensuing heat from radiating back to space. Based on the study on past climate shifts, notes of current situations, and computer simulations, many climate scientists say that lacking of big curbs in greenhouse gas discharges, the 21st century might see temperatures rise of about 3 to 8 degrees, climate patterns piercingly shift, ice sheets contract and seas rise several feet. With the probable exemption of one more world war, a huge asteroid, or a fatal plague, global warming may be the only most danger to our planet earth.

Global Warming Causes

As said, the major cause of global warming is the emission of green house gases like carbon dioxide, methane, nitrous oxide etc into the atmosphere. The major source of carbon dioxide is the power plants. These power plants emit large amounts of carbon dioxide produced from burning of fossil fuels for the purpose of electricity generation. About twenty percent of carbon dioxide emitted in the atmosphere comes from burning of gasoline in the engines of the vehicles. This is true for most of the developed countries. Buildings, both commercial and residential represent a larger source of global warming pollution than cars and trucks.

Building of these structures require a lot of fuel to be burnt which emits a large amount of carbon dioxide in the atmosphere. Methane is more than 20 times as effectual as CO_2 at entrapping heat in the atmosphere. Methane is obtained from resources such as rice paddies, bovine flatulence, bacteria in bogs and fossil fuel manufacture. When fields are flooded, anaerobic situation build up and the organic matter in the soil decays, releasing methane to the atmosphere. The main sources of nitrous oxide include nylon and nitric acid production, cars with catalytic converters, the use of fertilizers in agriculture and the burning of organic matter. Another cause of global warming is deforestation that is caused by cutting and burning of forests for the purpose of residence and industrialization.

Global Warming is Inspiring Scientists to Fight for Awareness

Scientists all over the world are making predictions about the ill effects of Global warming and connecting some of the events that have taken place in the pat few decades as an alarm of global warming. The effect of global warming is increasing the average temperature of the earth. A rise in earth's temperatures can in turn root to other alterations in the ecology, including an increasing sea level and modifying the quantity and pattern of rainfall. These modifications may boost the occurrence and concentration of severe climate events, such as floods, famines, heat waves, tornados, and twisters. Other consequences may comprise of higher or lower agricultural outputs, glacier melting, lesser summer stream flows, genus extinctions and rise in the ranges of disease vectors. As an effect of global warming species like golden toad, harlequin frog of Costa Rica has already become extinct. There are number of species that have a threat of disappearing soon as an effect of global warming. As an effect of global warming various new diseases have emerged lately. These diseases are occurring frequently due to the increase in earths average temperature since the bacteria can survive better in elevated temperatures and even multiplies faster when the conditions are favorable. The global warming is extending the distribution of mosquitoes due to the increase in humidity levels and their frequent growth in warmer atmosphere. Various diseases due to ebola, hanta and machupo virus are expected due to warmer climates. The marine life is also very sensitive to the increase in temperatures. The effect of global warming will definitely be seen on some species in the water. A survey was made in which the marine life reacted significantly to the changes in water temperatures. It is expected that many species will die off or become extinct due to the increase in the temperatures of the water, whereas various other species, which prefer warmer waters, will increase tremendously. Perhaps the most disturbing changes are expected in the coral reefs that are expected to die off as an effect of global warming. The global warming is expected to cause irreversible changes in the ecosystem and the behavior of animals.

A group of scientists have recently reported on the surprisingly speedy rise in the discharge of carbon and methane release from frozen tundra in Siberia, now starting to melt because of human cause increases in earth's temperature. The scientists tell us that the tundra is in danger of melting holds an amount of extra global warming pollution that is equivalent to the net amount that is previously in the earth's atmosphere. Likewise, earlier one more team of scientists reported that the in a single year Greenland witnessed 32 glacial earthquakes between 4.6 and 5.1 on the Richter scale. This is a disturbing sign and points that a huge destabilization that may now be in progress deep within the second biggest accretion of ice on the planet. This ice would be enough to raise sea level 20 feet worldwide if it broke up and slipped into the sea. Each day passing brings yet new proof that we are now in front of a global emergency, a climate emergency that needs instant action to piercingly decrease carbon dioxide emissions worldwide in order to turn down the earth's rising temperatures and avoid any catastrophe.

It is not easy to attach any particular events to global warming, but studies prove the fact that human activities are increasing the earth's temperature. Even though most predictions focus on the epoch up to 2100, even if no further greenhouse gases were discharged after this date, global warming and sea level would be likely to go on to rise for more than a millennium, since carbon dioxide has a long average atmospheric life span.

You Can Help Fight Global Warming

Many efforts are being made by various nations to cut down the rate of global warming. One such effort is the Kyoto agreement that has been made between various nations to reduce the emissions of various green house gases. Also many non profit organizations are working for the cause. Al Gore was one of the foremost U.S. politicians to heave an alarm about the hazards of global warming. He has produced a significantly acclaimed documentary movie called "An Inconvenient Truth," and written a book that archives his advice that Earth is dashing toward an immensely warmer future. Al Gore, the former vice president of United States has given various speeches to raise an awareness of global warming. He has warned people about the ill effects of Global warming and its remedies.

But an interesting side of the global warming episode is that there are people who do not consider global warming as something that is creating a problem. Skeptics of global warming think that global warming is not an ecological trouble. According to the global warming skeptics, the recent enhancement in the earth's average temperature is no reason for alarm. According to them earth's coastlines and polar ice caps are not at a risk of vanishing. Global warming skeptics consider that the weather models used to establish global warming and to forecast its impacts are distorted. According to the models, if calculations are made the last few decades must have been much worse as compared to actually happened to be. Most of the global warming skeptics believe that the global warming is not actually occurring. They stress on the fact the climatic conditions vary because of volcanism, the obliquity cycle, changes in solar output, and internal variability. Also the warming can be due to the variation in cloud cover, which in turn is responsible for the temperatures on the earth. The variations are also a result of cosmic ray flux that is modulated by the solar magnetic cycles.

Global Warming Skeptics

The global warming skeptics are of the view that the global warming is a good phenomenon and should not be stopped. There are various benefits of global warming according to them. According to the skeptics, the global warming will increase humidity in tropical deserts. Also the higher levels of carbon dioxide in the atmosphere trigger plant growth. As predicted, due to the global warming the sea levels will rise. But this can be readily adapted. Another argument of global warming skeptics is that earth has been warmer than today as seen in its history. The thought is that global warming is nothing to get afraid of because it just takes us back to a more natural set of environment of the past. Animals and plants appeared to do just fine in those eras of warm climate on the earth. According to few skeptics, the present chilly climate on the earth is an abnormality when judged over the geographical scale. Over geologic time, the earth's mean temperature is 22 degrees C, as compared to today's 15.5 degrees C.

Ways to Prevent Global Warming

Global warming is essentially the excessive heating of the planet, which results in dramatic climatic changes. So, are there ways to prevent global warming?

In many ways, the global warming debate is benefiting from the current highly volatile issues concerning energy. The world wide industrial revolution has been powered by coal and oil. While coal is available in many countries, oil is not. Circumstances being what they are, the largest concentrations of oil are found in areas with significant political and armed conflict. One needs only look at the situation in Iraq to understand this fact. With the growing China economy, there now seems to be a universal understanding that oil alternatives are necessary.

Perhaps proving the concept of Karma, nearly all oil alternatives are favorable ways to reduce global warming. Whether we are talking about hydropower, solar, wind, geothermal or even nuclear energy, each alternative power source produces significantly less pollution than classic fossil fuels. Less pollution is the key for one very simple reason.

A major byproduct of using fossil fuels for energy is the much discussed greenhouse gas. Whether using oil to produce electricity or just to gas up the car, the use of this medium produces carbon dioxide. Carbon dioxide is a major greenhouse gas. Reducing the volume of it in the atmosphere will greatly reduce the greenhouse effect. As our energy production evolves away from oil, we should see improvements in the atmosphere. The bigger issue, of course, is whether we will get around to this anytime soon.

A second major way to prevent global warming is to stop deforestation. Vast forests, such as the one found in the Amazon River Basin in South America are being clear cut. This has a devastating impact on global warming. Nature being what it is, it has a natural method for filtering harmful gases such as carbon dioxide out of the air. It does this through photosynthesis with plants. Dense, huge forests are the biggest and best filters of carbon dioxide. By cutting them down, we reduce the ability of the planet to filter out the bad gases while we continue to pump more of the gases into the atmosphere. Any way you measure it, it is a formula destined to cause trouble.

Are there ways to prevent global warming? Yes. We need to kick our oil habit and stop wiping out our forest areas.

Four Easy Ways To Help Prevent Global Warming: Start With Simple Steps Anyone Can Take

You want to know how you can easily help prevent global warming?

There are actually many ways and I've included four below that are simple and easy for you to implement right away. All four of these steps focus on reducing your energy consumption in order to reduce the amount of greenhouse gases you are personally responsible for generating. When you look at this list you may think to yourself that these things don't have a big impact. If that's what you think, then you need to understand...

If you only pay attention to your own actions, then you are not going to be effective in preventing global warming!

If you are the only person to implement the following recommendations, you'd be right; your actions wouldn't have a big impact. But, by making these simple changes you show the people around you ways to prevent global warming by leading by example.

By being the change you wish to see, you greatly increase your credibility in the eyes of your peers, which makes it much easier for you to enroll them in helping to prevent global warming with you.

So, don't underestimate the power of these four easy ways to prevent global warming!

Step 1: Prevent global warming by minimizing drafts in your home.

One of the simplest and easiest ways to prevent global

warming is to make sure all your windows and doors are draft free. Small gaps in your windows and doors can cause you to consume a lot more energy than you really need to in order to heat and cool your home. To make your home free from drafts wait until a cold day, then hold your hand along the edges of your doors and windows feeling for cold air. If you detect drafts, buy weather stripping from your local hardware store and install it wherever it is needed.

Step 2: Prevent global warming by reducing your waste electricity use.

Amazingly about 11% of residential electricity consumption is used by "phantom loads." A phantom load is caused when an electrical appliance draws electricity when it isn't in use. For example, a DVD player has an electrically lighted display that stays on even when you turn the power off. An easy way to eliminate these phantom loads is to plug your computers, printers, scanners, DVD players, televisions, etc. into multi-plug electrical surge protectors. Then, with one flick of the switch, you can make sure that the appliances plugged into your surge protectors are drawing no electricity at all.

Step 3: Prevent global warming with more efficient light bulbs!

One of the simplest ways to prevent global warming is by using more efficient light bulbs. The Union of Concerned Scientists tell us that if every American household replaced just one incandescent light bulb with a compact fluorescent one, then over the life of the bulbs 90 billion pounds of CO2 emission would be prevented from going into the atmosphere. This is like taking 6.3 million cars off the road!

Step 4: Prevent global warming by turning down your hot water heater.

Most of us have our hot water heaters turned up too high. If you turn down your hot water thermostat just 10 degrees Fahrenheit, you reduce 3% to 5% of your energy consumption. You'll have plenty of hot water if you set your water heater between 110-120°F.

These four simple ways to prevent global warming can be implemented quickly, easily, and with little to no expense.

If the potential effects of global warming are really scary to you, then you should implement these steps in your life today, but don't stop there!

Once you've implemented these solutions, share this article with your friends, and explain to your friends how you've made these simple changes. Be sure to offer your friends help and support in making these changes themselves to make it extra easy for them to prevent global warming with you.

Also, when you tell your friends about this article and the changes you've made, be sure to explain all the benefits you've received from doing so. Help your friends feel how these four simple steps have helped give you more hope, save you money, etc! The more your friends can feel the emotional and financial benefits you've experienced from preventing global warming, they more likely they are to join you!

Lead by example, and help others to do the same! The cumulative effect will be significant if you put dedicated effort into changing yourself while supporting those around you to do the same. So don't delay, be the change you wish to see. Implement these four setps so you can help in preventing global warming today, and then get others to join you in your efforts!

4

Recent Developments of Climate Change

This draft policy brief says African, Caribbean and Pacific (ACP) countries can use local natural resources — such as sugarcane and *jatropha* — to meet energy requirements through biofuels, curtailing dependence on fossil fuels.

But the authors warn of challenges for developing countries, including economic and trade issues, and suggest practical steps for meeting these. They also present various bioenergy options for households, such as BioGel — a solid wood-substitute made from low-grade ethanol mixed with a gelling agent.

The brief makes a number of policy recommendations, including national strategies for promoting and sustaining local demand, and more funding for local and regional ACP research.

This document summarises the outcomes from the twelfth UN Conference of the Parties (COP) to the Framework Convention on Climate Change, and the concurrent Meeting of the Parties to the Kyoto Protocol in Nairobi, Kenya in November 2006.

Parties to the conference discussed the adaptation fund, deciding on a governance mechanism for it, but they sidestepped the issue of where it should be administered. They also considered how to promote Clean Development Mechanism projects in developing countries, ultimately stressing the need for more of these in Africa.

Wide debate on the mandate of the Expert Group on Technology Transfer led to a decision to reassess the group's work next year. Discussions on how to use incentives to 'avoid' deforestation in developing countries and reduce emissions also proved difficult to reconcile.

The summary is an accessible and authoritative guide to recent events at the climate negotiations, even for those unfamiliar with the UN climate change process.

Climate change mitigation in developing countries

The role of developing countries in climate change mitigation has been and continues to be a contentious issue. Developing countries' emissions are predicted to surpass those of industrialised countries within the first half of this century, but no formal commitments to reduce emissions have been made.

This report, prepared for the Pew Center on Global Climate Change, examines six countries — Brazil, China, India, Mexico, South Africa and Turkey — in the context of climate change mitigation. Ongoing efforts in these countries have helped reduce emissions, though not necessarily in the name of mitigating climate change.

The authors find that overall, over the past three decades, these countries have reduced the growth rate of their emissions by 300 million tonnes. The motivations for such efforts include poverty alleviation, economic development, energy security and local environmental protection. This demonstrates that climate change mitigation can and does occur in the context of development that aims to be sustainable.

This report is comprehensive for the countries studied. It is very accessible and likely to be of interest to anyone engaged in the debate about mitigation in the South.

The report is available in pdf format only. An executive summary is availably online here.

Climate Change: India's Perceptions, Positions, Policies and Possibilities

The threat of climate change is perceived differently by different countries. This informal working paper, commissioned by the Organisation for Economic Co-operation and Development (OECD), comments on the key interactions between climate change and sustainable development in India. After outlining the risks climate change poses to India, it describes some of the national efforts that have been taken to reduce greenhouse gas emissions, including the promotion of renewable energy, carbon sequestration through afforestation programmes and price reforms for improving energy efficiency. The final section considers the options available to India for mitigating climate change and includes a cost-benefit analysis.

Climate Policy Beyond 2012

Source: Centre for International Climate and Environmental Research (CICERO), May 2004

The first commitment period of the Kyoto Protocol on climate change ends in 2012; at the time of writing, it remains unclear what will follow. Various approaches have been suggested, and the authors of this report analyse current thinking on future climate policy and make their own recommendations.

The report considers long-term climate policy targets, climate policy frameworks and their architecture, issues related to adaptation and sustainable development.

The major challenges, issues and questions concerning the design of future climate policy are addressed throughout.

While somewhat lengthy, the report provides a thorough theoretical background for anyone interested in the intricacies of future climate policy. It is complementary reading to the Pew Center report *International Climate Efforts Beyond 2012* .

International Climate Efforts beyond 2012: A survey of approaches

Source: Pew Center on Global Climate Change, December 2004

The emerging discussion on international climate policy after 2012 (the end of the first commitment period of the Kyoto Protocol) has generated a number of approaches. Many of these are only now being analysed.

This report provides a comprehensive yet succinct overview of 43 different approaches to international climate efforts. Following an overview of key issues, each approach is explained in terms of its rationale, forum, time frame, mitigation commitment, institutional arrangements and other elements.

The document provides a reference guide to the essential characteristics of post-2012 climate approaches. It is most useful, and key reading for anyone interested in climate policy.

A guide to the climate change convention and its Kyoto protocol

Source: United Nations Framework Convention on Climate Change (UNFCCC), May 2002

On 9 May 1992, the world's governments adopted the UN Framework Convention on Climate Change. Five years later, on 11 December 1997, governments took a further step forwards and adopted the landmark Kyoto Protocol.

Building on the framework of the Convention, the Kyoto Protocol broke new ground with its legally-binding constraints on greenhouse gas emissions and its innovative "mechanisms" aimed at cutting the cost of curbing emissions. Today, 186 countries (including the European Community) are Parties to the Convention, more than most any other environmental treaty, and the entry into force of the Kyoto Protocol is expected soon.

This guide, prepared in the tenth anniversary year of the adoption of the Convention, explains in detail the commitments of both the Convention and the Kyoto Protocol, along with the "rulebook" for their implementation.

This is the full text of the Framework Convention, which was adopted at the United Nations Headquarters, New York on 9 May 1992. The convention was open for signature at the Earth Summit in Rio de Janeiro from 4 to 14 June 1992, and thereafter at the UN Headquarters in New York, from 20 June 1992 to 19 June 1993. By that date the Convention had received 166 signatures. The Convention entered into force on 21 March 1994.

International emissions trading and global climate change

Source: Pew Center on Global Climate Change, December 1999

Several factors influence the costs of greenhouse gas mitigation. This report illustrates the importance of one such factor — international emissions trading — in reducing the costs of carbon control. The authors argue that an international greenhouse gas emissions trading regime will significantly lower global mitigation costs.

The Third Assessment Report of the IPCC's Working

Group 1 builds on past assessments and incorporates new results from the past five years of climate change. It descibes the current state of udnerstanding of the cliamte system, and provides estiamtes of its projected future evolution and their uncertainties.

Many hundreds of scientists from around the world participated in the preparation and review of the report, which states that "there is new and stronger evidence that most of the warming observed over the last 50 years is attributable to human activities".

Climate change in focus: the IPCC Third Assessment Report. The Third Assessment Report of the Intergovernmental Panel on Climate Change, published in 2001, is the most comprehensive and authoritative source of information on climate change. Its conclusions confirm and strengthen those of the previous reports: human-induced climate change is a reality and most of the effects will be negative, but a range of mitigation opportunities is available to address the problem.

The Report finds that most of the earth's warming over the past 50 years can be attributed to human activities, and that its effects are already being felt. Global temperature is expected to increase by 1.4 to 5.8ºC over the next century, a significant increase on the projections of the 1995 Second Assessment Report. This briefing paper summarises the findings of the Third Assessment Report and the debates underpinning them, and discusses the likely outcomes of the Report.

Recent Developments

The first High-level Biofuels Seminar in Africa, which met in Addis Ababa, Ethiopia, from 30 July to 1 August 2007, has developed an Action Plan for Biofuels Development in Africa. The Action Plan was annexed to the Addis

Ababa Declaration on Su.

World Water Week opens in Stockholm amid biofuel demand warnings. Top experts from 140 countries gathered in Stockholm for a week-long international conference on water sustainability initiatives, amid warnings that expected mass production of biofuels such as ethanol for the automobile industry may exacerbate.

Jamaica: Castor beans for bio-diesel production - Enthusiasts propose mass cultivation to cut fuel cost

Biofuel enthusiasts here are proposing the mass cultivation of castor beans - a shrub that already grows in the wild across the island - for the production of bio-diesel to be used as a fuel blend in the transport sector.

Guyana poised for significant biofuel future

At the high-level seminar on expanding bio-energy opportunities in the Caribbean, Prime Minister Samuel Hinds expressed optimism about Guyana's pursuit of that new sector. He referred to numerous plans for investment from stakeholders seeking le.

Canadian firm to invest $55 million in Mozambique jatropha venture. Energem Resources of Canada says that it has acquired a 70 per cent stake of shares in a renewable energy venture in Mozambique that will use jatropha to produce biodiesel fuel. The firm said that under the terms of its purchase, it will invest.

UK firm invests $20m in Tanzania biofuel venture

Tanzania has landed a $20 million biofuel processing project that will see large-scale planting of jatropha for the production and distribution of crude and refined products.

Spanish company explores jatropha cultivation, processing in Ghana

Spanish company Entaban S.A. has expressed interest in establishing a jatropha plantation and bio-diesel plant at Offinso in the Ashanti Region of Ghana. An invitation has been issued by the company for government officials to visit the company'...

South Africa: Growing fuel

Few would argue with Science and Technology Minister Mosibudi Mangena's insistence that we tread carefully when developing a national biofuels policy. At the same time, it would be folly to delay the legislation on which such an industry.

Chinese firm to invest $1 billion in 3 million hectare DR Congo oil palm plantation. A Chinese company, ZTE International, is to invest US$1 billion in a 3 million hectare oil palm plantation in the Democratic Republic of Congo (DRC) to produce biofuels. The vast Central-African country is a potential bioenergy 'superpower'

EU aid to help developing countries jump on biofuel bandwagon. The EU plans to give money to help developing countries grow energy crops rich nations want to use for transport fuel, says EU Development Commissioner Louis Michel. Europe and the United States plan to use more low-carbon emission biofuels.

Global Bioenergy Partnership website launched

The Global Bioenergy Partnership (GBEP) unveiled its website, which provides information on the Partnership, created in May 2006 to promote the use of bioenergy the secretariat of which is hosted at FAO. The website also offers links to sources.

EU launches platform to promote bioenergy in semi-arid Africa

The EU has launched an ambitious project to study

and implement bioenergy projects in the semi-arid and arid regions of Africa, aimed at bringing modern biofuels to local populations, as well as exporting green fuels to world markets.

Entrepreneurs target the jatropha plant as the next big biofuel A woody shrub with big oily seeds could be the ideal source for biofuel. For hundreds of years, Africans have used Jatropha curcas (jatropha) as a living fence. Now biodiesel entrepreneurs in Africa and India are buying up land, starting plantat.

Study questions benefits of biofuels

Biofuels may not be the panacea for the world's fossil-fuel woes, according to a new study. Such fuels, touted as an ecologically friendly source of energy, may be more harmful for the environment than their fossil counterparts.

South Africa: Farmers looking for a lift from 'ethanol effect' Agricultural economists call it the ethanol effect: an increase in world maize prices based on an increase in the demand for energy, which they say will ensure the future of agriculture in SA. Although the government is incorporating biofuels.

A large-scale biogas initiative has been launched to bring renewable energy to 20 million households in some 25 African countries. The initiative was approved at a conference entitled 'Biogas for Better Life: An African Initiative', held in Nair.

The National Institute of Scientific Research and Technology (IRST) will test the environment alternative source of power that uses a combination of plant husks and clay. The energy source, once scientifically proved according to IRST researchers.

Speaking in Nairobi, Kenya, at the fourth monthly

Open Forum for Agricultural Biotechnology (OFAB 4), UNIDO Director General Kandeh Yumkellah called on African scientists and leaders to make Africa a world leader in biofuels production.

Research for development: A Global Issue

The supplement presents views from 30 scientists on key topics in this area of research, which is central to global issues including sustainable development, climate change, poverty, animal health, globalization, biofuels, water and biodiversity.

Making biofuels without wasting food

An EU Strategy for Biofuels

EC, 8 February 2006.The EU is supporting biofuels with the aim of reducing greenhouse gas emissions, boosting the decarbonisation of transport fuels, diversifying fuel supply sources, offering new income opportunities in rural areas and developin.

Global warming is the increase in the average temperature of the Earth's near-surface air and oceans since the mid-twentieth century and its projected continuation.The average global air temperature near the Earth's surface increased 0.74 ± 0.18 °C (1.33 ± 0.32 °F) during the hundred years ending in 2005.[1] The Intergovernmental Panel on Climate Change (IPCC) concludes "most of the observed increase in globally averaged temperatures since the mid-twentieth century is very likely due to the observed increase in anthropogenic greenhouse gas concentrations"[1] via the greenhouse effect. Natural phenomena such as solar variation combined with volcanoes probably had a small warming effect from pre-industrial times to 1950 and a small cooling effect from 1950 onward.[2][3] These basic conclusions have been endorsed by at least thirty scientific societies and academies of science,[4] including all of the national academies of science of the major industrialized countries.[5][6][7] While individual scientists have voiced disagreement with some findings of the IPCC,[8] the overwhelming majority of scientists working on climate change agree with the IPCC's main conclusions.[9][10] Climate model projections summarized by the IPCC indicate that average global surface temperature will likely rise a further 1.1 to 6.4 °C (2.0 to 11.5 °F) during the twenty-first century.[1] The range of values results from the use of differing scenarios of future greenhouse gas emissions as well as models with differing climate sensitivity. Although most studies focus on the period up to 2100, warming and sea level rise are expected to continue for more than a thousand years even if greenhouse gas levels are stabilized. The delay in reaching equilibrium is a result of the large heat capacity of the oceans.[1]Increasing global temperature will cause sea level to rise, and is expected to increase the intensity of extreme weather events and to change the amount and pattern of precipitation. Other effects of global warming include changes in agricultural yields, trade routes, glacier retreat, species extinctions and increases in the ranges of disease vectors.Remaining scientific uncertainties include the amount of warming expected in the future, and how warming and related changes will vary from region to region around the globe. Most national governments have signed and ratified the Kyoto Protocol aimed at reducing greenhouse gas emissions, but there is ongoing political and public debate worldwide regarding what, if any, action should be taken to reduce or reverse future warming or to adapt to its expected consequences.

Terminology

The term "global warming" is a specific example of global climate change. The term "climate change" can also refer to other periods of overall temperature change such as global cooling. In common usage, the term "global warming" refers to the warming in recent decades and its projected continuation, and implies a human influence.[11][12] The United Nations Framework Convention on Climate Change (UNFCCC) uses the term "climate change" for human-caused change, and "climate variability" for other changes.[13] The term "anthropogenic global warming" (AGW) is sometimes used when focusing on human-induced changes.

Causes Components of the current radiative forcing as estimated by the IPCC Fourth Assessment Report.*Main articles: Attribution of recent climate change and Scientific opinion on climate change*The Earth's climate changes in response to external forcing, including variations in its orbit around the Sun (orbital forcing),[14][15][16] volcanic eruptions,[17] and atmospheric greenhouse gas concentrations. The detailed causes of the recent warming remain an active field of research, but the scientific consensus[18][19] is that the increase in atmospheric greenhouse gases due to human activity caused most of the warming observed since the start of the industrial era. This attribution is clearest for the most recent 50 years, for which the most detailed data are available. Some other hypotheses departing from the consensus view have been suggested to explain most of the temperature increase. One such hypothesis proposes that warming may be the result of variations in solar activity.[20][21][22]None of the effects of forcing are instantaneous. The thermal inertia of the Earth's oceans and slow responses of other indirect effects mean that the Earth's current climate is not in equilibrium with the forcing imposed. Climate commitment studies indicate that even if greenhouse gases were stabilized at 2000 levels, a further warming of about 0.5 °C (0.9 °F) would still occur.[23]

Greenhouse gases in the atmosphere. *Greenhouse gas and Greenhouse effect* The greenhouse effect was discovered by Joseph Fourier in 1824 and was first investigated quantitatively by Svante Arrhenius in 1896. It is the process by which absorption and emission of infrared radiation by atmospheric gases warm a planet's lower atmosphere and surface.Existence of the greenhouse effect as such is not disputed. Naturally occurring greenhouse gases have a mean warming effect of about 33 °C (59 °F), without which Earth would be uninhabitable.[24][25] On Earth, the major greenhouse gases are water vapor, which causes about 36–70% of the greenhouse effect (not including clouds); carbon dioxide (CO_2), which causes 9–26%; methane (CH_4), which causes 4–9%; and ozone, which causes 3–7%.[26][27] The issue is how the strength of the greenhouse effect changes when human activity increases the atmospheric concentrations of some greenhouse gases.Human activity since the industrial revolution has increased the concentration of various greenhouse gases, leading to increased radiative forcing from CO_2, methane, tropospheric ozone, CFCs and nitrous oxide. Molecule for molecule, methane is a more effective greenhouse gas than carbon dioxide, but its concentration is much smaller so that its total radiative forcing is only about a fourth of that from carbon dioxide. Some other naturally occurring gases contribute very small fractions of the greenhouse effect; one of these, nitrous oxide (N_2O), is increasing in concentration owing to human activity such as agriculture. The atmospheric concentrations of CO_2 and CH_4 have increased by 31% and 149% respectively since the beginning of the industrial revolution in the mid-1700s. These levels are considerably higher than at any time during the last 650,000 years, the period for which reliable data has been extracted from ice cores.[28] From less direct geological evidence it is believed that CO_2 values this high were last attained 20 million years ago.[29] Fossil fuel burning has produced about three-quarters of the increase in CO_2 from human activity over the past 20 years. Most of the rest is due to land-use change, in particular deforestation.[30]Recent increases in atmospheric carbon dioxide (CO_2). The monthly CO_2 measurements display small seasonal oscillations in an overall yearly uptrend; each year's maximum is reached during the Northern Hemisphere's late spring, and declines during the Northern Hemisphere growing season as plants remove some CO_2 from the atmosphere.The present atmospheric concentration of CO_2 is about 385 parts per million (ppm) by volume.[31] Future CO_2 levels are expected to rise due to ongoing burning of fossil fuels and land-use change. The rate of rise will depend on uncertain economic, sociological, technological, and natural developments, but may be ultimately limited by the availability of fossil fuels. The IPCC Special Report on Emissions Scenarios gives a wide range of future CO_2 scenarios, ranging from 541 to 970 ppm by the year 2100.[32] Fossil fuel reserves are sufficient to reach this level and continue emissions past 2100, if coal, tar sands or methane clathrates are extensively used.[33]

*Feedbacks Effects of global warming Positive feedback effects*The effects of forcing agents on the climate are complicated by various feedback processes.One of the most pronounced feedback effects relates to the evaporation of water. Warming by the addition of long-lived greenhouse gases such as CO_2 will cause more water to evaporate into the atmosphere. Since water vapor itself acts as a greenhouse gas, the atmosphere warms further; this warming causes more water vapor to evaporate (a positive feedback), and so on until other processes stop the feedback loop. The result is a much larger greenhouse effect than that due to CO_2 alone. Although this feedback process causes an increase in the absolute moisture content of the air, the relative humidity stays nearly constant or even decreases slightly because the air is warmer.[34] This feedback effect can only be reversed slowly as CO_2 has a long average atmospheric lifetime.Feedback effects due to clouds are an area of ongoing research. Seen from below, clouds emit infrared radiation back to the surface, and so exert a warming effect; seen from above, clouds reflect sunlight and emit infrared radiation to space, and so exert a cooling effect. Whether the net effect is warming or cooling depends on details such as the type and altitude of the cloud. These details are difficult to represent in climate models, in part because clouds are much smaller than the spacing between points on the computational grids of climate models. Nevertheless, cloud feedback is second only to water vapor feedback and is positive in all the models that were used in the IPCC Fourth Assessment Report.[34]A subtler feedback process relates to changes in the lapse rate as the atmosphere warms. The atmosphere's temperature decreases with height in the troposphere. Since emission of infrared radiation varies with the fourth power of temperature, longwave radiation emitted from the upper atmosphere is less than that emitted from the lower atmosphere. Most of the radiation emitted from the upper atmosphere escapes to space, while most of the radiation emitted from the lower atmosphere is re-absorbed by the surface or the atmosphere. Thus, the strength of the greenhouse effect depends on the atmosphere's rate of temperature decrease with height: if the rate of temperature decrease is greater the greenhouse effect will be stronger, and if the rate of temperature decrease is smaller then the greenhouse effect will be weaker. Both theory and climate models indicate that warming will reduce the decrease of temperature with height, producing a negative *lapse rate feedback* that weakens the greenhouse effect. Measurements of the rate of temperature change with height are very sensitive to small errors in observations, making it difficult to establish whether the models agree with observations.[35]Another important feedback process is ice-albedo feedback.[36] When global temperatures increase, ice near the poles melts at an increasing rate. As the ice melts, land or open water takes its place. Both land and open water are on average less reflective than ice, and thus absorb more solar radiation.

This causes more warming, which in turn causes more melting, and this cycle continues.Positive feedback due to release of CO_2 and CH_4 from thawing permafrost, such as the frozen peat bogs in Siberia, is an additional mechanism that could contribute to warming.[37] Similarly a massive release of CH_4 from methane clathrates in the ocean could cause rapid warming, according to the clathrate gun hypothesis.The ocean's ability to sequester carbon is expected to decline as it warms. This is because the resulting low nutrient levels of the mesopelagic zone (about 200 to 1000 m depth) limits the growth of diatoms in favor of smaller phytoplankton that are poorer biological pumps of carbon.[38]

Solar variation. Solar variation over the last thirty years.*Main article: Solar variation*A few papers suggest that the Sun's contribution may have been underestimated. Two researchers at Duke University, Bruce West and Nicola Scafetta, have estimated that the Sun may have contributed about 45–50% of the increase in the average global surface temperature over the period 1900–2000, and about 25–35% between 1980 and 2000.[39] A paper by Peter Stott and other researchers suggests that climate models overestimate the relative effect of greenhouse gases compared to solar forcing; they also suggest that the cooling effects of volcanic dust and sulfate aerosols have been underestimated.[40] They nevertheless conclude that even with an enhanced climate sensitivity to solar forcing, most of the warming since the mid-20th century is likely attributable to the increases in greenhouse gases.A different hypothesis is that variations in solar output, possibly amplified by cloud seeding via galactic cosmic rays, may have contributed to recent warming.[41] It suggests magnetic activity of the sun is a crucial factor which deflects cosmic rays that may influence the generation of cloud condensation nuclei and thereby affect the climate.[42]One predicted effect of an increase in solar activity would be a warming of most of the stratosphere, whereas greenhouse gas theory predicts cooling there.[43] The observed trend since at least 1960 has been a cooling of the lower stratosphere.[44] Reduction of stratospheric ozone also has a cooling influence, but substantial ozone depletion did not occur until the late 1970s.[45] Solar variation combined with changes in volcanic activity probably did have a warming effect from pre-industrial times to 1950, but a cooling effect since.[1] In 2006, Peter Foukal and other researchers from the United States, Germany, and Switzerland found no net increase of solar brightness over the last thousand years. Solar cycles led to a small increase of 0.07% in brightness over the last thirty years. This effect is far too small to contribute significantly to global warming.[46][47] A paper by Mike Lockwood and Claus Fröhlich found no relation between global warming and solar radiation since 1985, whether through variations in solar output or variations in cosmic rays.[48] Henrik Svensmark and Eigil Friis-Christensen, the main proponents of cloud seeding by galactic cosmic rays, disputed this criticism of their hypothesis.[49] A 2007 paper found that in the last 20 years there has been no significant link between changes in cosmic rays coming to Earth and cloudiness and temperature.[50][51][52]

Temperature changes Two millennia of mean surface temperatures according to different reconstructions, each smoothed on a decadal scale. The unsmoothed, annual value for 2004 is also plotted for reference.*Main article: Temperature record* Global temperatures on both land and sea have increased by 0.75 °C (1.35 °F) relative to the period 1860–1900, according to the instrumental temperature record. This measured temperature increase is not significantly affected by the urban heat island effect.[53] Since 1979, land temperatures have increased about twice as fast as ocean temperatures (0.25 °C per decade against 0.13 °C per decade).[54] Temperatures in the lower troposphere have increased between 0.12 and 0.22 °C (0.22 and 0.4 °F) per decade since 1979, according to satellite temperature measurements. Temperature is believed to have been relatively stable over the one or two thousand years before 1850, with possibly regional fluctuations such as the Medieval Warm Period or the Little Ice Age.Sea temperatures increase more slowly than those on land both because of the larger effective heat capacity of the oceans and because the ocean can lose heat by evaporation more readily than the land.[55] The Northern Hemisphere has more land than the Southern Hemisphere, so it warms faster. The Northern Hemisphere also has extensive areas of seasonal snow and sea-ice cover subject to the ice-albedo feedback. More greenhouse gases are emitted in the Northern than Southern Hemisphere, but this does not contribute to the difference in warming because the major greenhouse gases persist long enough to mix between hemispheres.Based on estimates by NASA's Goddard Institute for Space Studies, 2005 was the warmest year since reliable, widespread instrumental measurements became available in the late 1800s, exceeding the previous record set in 1998 by a few hundredths of a degree.[56] Estimates prepared by the World Meteorological Organization and the Climatic Research Unit concluded that 2005 was the second warmest year, behind 1998.[57][58] Temperatures in 1998 were unusually warm because the strongest El Niño in the past century occurred during that year.[59]Anthropogenic emissions of other pollutants—notably sulfate aerosols—can exert a cooling effect by increasing the reflection of incoming sunlight. This partially accounts for the cooling seen in the temperature record in the middle of the twentieth century,[60] though the cooling may also be due in part to natural variability. James Hansen and colleagues have proposed that the effects of the products of fossil fuel combustion—CO_2 and aerosols—have largely offset one another, so that warming in recent decades has been driven mainly by non-CO_2 greenhouse gases.[61]Paleoclimatologist William Ruddiman has argued that human influence on the global climate began around

8,000 years ago with the start of forest clearing to provide land for agriculture and 5,000 years ago with the start of Asian rice irrigation.[62] Ruddiman's interpretation of the historical record, with respect to the methane data, has been disputed.[63]

Pre-human climate variations Curves of reconstructed temperature at two locations in Antarctica and a global record of variations in glacial ice volume. Today's date is on the left side of the graph.*Further information: Paleoclimatology See also: Snowball Earth* Earth has experienced warming and cooling many times in the past. The recent Antarctic EPICA ice core spans 800,000 years, including eight glacial cycles timed by orbital variations with interglacial warm periods comparable to present temperatures.[64]A rapid buildup of greenhouse gases amplified warming in the early Jurassic period (about 180 million years ago), with average temperatures rising by 5 °C (9 °F). Research by the Open University indicates that the warming caused the rate of rock weathering to increase by 400%. As such weathering locks away carbon in calcite and dolomite, CO_2 levels dropped back to normal over roughly the next 150,000 years.[65][66]Sudden releases of methane from clathrate compounds (the clathrate gun hypothesis) have been hypothesized as both a cause for and an effect of other warming events in the distant past, including the Permian–Triassic extinction event (about 251 million years ago) and the Paleocene–Eocene Thermal Maximum (about 55 million years ago).

Calculations of global warming prepared in or before 2001 from a range of climate models under the SRES A2 emissions scenario, which assumes no action is taken to reduce emissions.The geographic distribution of surface warming during the 21st century calculated by the HadCM3 climate model if a business as usual scenario is assumed for economic growth and greenhouse gas emissions. In this figure, the globally averaged warming corresponds to 3.0 °C (5.4 °F).*Main article: Global climate model*Scientists have studied global warming with computer models of the climate. These models are based on physical principles of fluid dynamics, radiative transfer, and other processes, with simplifications being necessary because of limitations in computer power and the complexity of the climate system. All modern climate models include an atmospheric model that is coupled to an ocean model and models for ice cover on land and sea. Some models also include treatments of chemical and biological processes.[67]These models predict that the effect of adding greenhouse gases is to produce a warmer climate.[68] However, even when the same assumptions of future greenhouse gas levels are used, there still remains a considerable range of climate sensitivity.Including uncertainties in future greenhouse gas concentrations and climate modeling, the IPCC anticipates a warming of 1.1 °C to 6.4 °C (2.0 °F to 11.5 °F) by the end of the 21st century, relative to 1980–1999.[1] Models have also been used to help investigate the causes of recent climate change by comparing the observed changes to those that the models project from various natural and human-derived causes.Current climate models produce a good match to observations of global temperature changes over the last century, but do not simulate all aspects of climate.[69] These models do not unambiguously attribute the warming that occurred from approximately 1910 to 1945 to either natural variation or human effects; however, they suggest that the warming since 1975 is dominated by man-made greenhouse gas emissions.Global climate model projections of future climate are forced by imposed greenhouse gas emission scenarios, most often from the IPCC Special Report on Emissions Scenarios (SRES). Less commonly, models may also include a simulation of the carbon cycle; this generally shows a positive feedback, though this response is uncertain (under the A2 SRES scenario, responses vary between an extra 20 and 200 ppm of CO_2). Some observational studies also show a positive feedback.[70][71][72]The representation of clouds is one of the main sources of uncertainty in present-generation models, though progress is being made on this problem.[73]

Sparse records indicate that glaciers have been retreating since the early 1800s. In the 1950s measurements began that allow the monitoring of glacial mass balance, reported to the WGMS and the NSIDC.

Effects of global warming. Although it is difficult to connect specific weather events to global warming, an increase in global temperatures may in turn cause broader changes, including glacial retreat, Arctic shrinkage, and worldwide sea level rise. Changes in the amount and pattern of precipitation may result in flooding and drought. There may also be changes in the frequency and intensity of extreme weather events. Other effects may include changes in agricultural yields, addition of new trade routes,[74] reduced summer streamflows, species extinctions, and increases in the range of disease vectors. Some effects on both the natural environment and human life are, at least in part, already being attributed to global warming. A 2001 report by the IPCC suggests that glacier retreat, ice shelf disruption such as that of the Larsen Ice Shelf, sea level rise, changes in rainfall patterns, and increased intensity and frequency of extreme weather events, are being attributed in part to global warming.[75] While changes are expected for overall patterns, intensity, and frequencies, it is difficult to attribute specific events to global warming. Other expected effects include water scarcity in some regions and increased precipitation in others, changes in mountain snowpack, and adverse health effects from warmer temperatures.[76] Increasing deaths, displacements, and economic losses projected due to extreme weather attributed to global warming may be exacerbated by growing population densities in affected areas, although temperate regions are projected to experience some benefits, such as fewer deaths due to cold exposure.[77] A summary of probable effects and recent understanding can be found in the report made for the IPCC Third Assessment Report by Working Group II.[75] The newer IPCC Fourth Assessment Report summary reports that there is observational evidence for an increase in intense tropical cyclone activity in the North Atlantic Ocean since about 1970, in correlation with the increase in sea surface temperature, but that the detection of long-term trends is complicated by the quality of records prior to routine satellite observations. The summary also states that there is no clear trend in the annual worldwide number of tropical cyclones.[1]Additional anticipated effects include sea level rise of 110 to 770 millimeters (0.36 to 2.5 ft) between 1990 and 2100,[78] repercussions to agriculture, possible slowing of the thermohaline circulation, reductions in the ozone layer, increased intensity of hurricanes and extreme weather events, lowering of ocean pH, and the spread of diseases such as malaria and dengue fever. One study predicts 18% to 35% of a sample of 1,103 animal and plant species would be extinct by 2050, based on future climate projections.[79] However, few mechanistic studies have documented extinctions due to recent climate change[80] and one study suggests that projected rates of extinction are uncertain.[81]

The projected temperature increase for a range of stabilization scenarios (the colored bands). The black line in middle of the shaded area indicates 'best estimates'; the red and the blue lines the likely limits. From the work of IPCC AR4.

Economics of global warming and Low-carbon economy
Some economists have tried to estimate the aggregate net economic costs of damages from climate change across the globe. Such estimates have so far failed to reach conclusive

findings; in a survey of 100 estimates, the values ran from US$-10 per tonne of carbon (tC) (US$-3 per tonne of carbon dioxide) up to US$350/tC (US$95 per tonne of carbon dioxide), with a mean of US$43 per tonne of carbon (US$12 per tonne of carbon dioxide).[77] One widely publicized report on potential economic impact is the Stern Review; it suggests that extreme weather might reduce global gross domestic product by up to 1%, and that in a worst-case scenario global per capita consumption could fall 20%.[82] The report's methodology, advocacy and conclusions have been criticized by many economists, primarily around the Review's assumptions of discounting and its choices of scenarios,[83] while others have supported the general attempt to quantify economic risk, even if not the specific numbers.[84][85]Preliminary studies suggest that costs and benefits of mitigation are broadly comparable in magnitude.[86]According to United Nations Environment Programme (UNEP), economic sectors likely to face difficulties related to climate change include banks, agriculture, transport and others.[87] Developing countries dependent upon agriculture will be particularly harmed by global warming.[88]

In November 2007, the Center for Strategic and International Studies and the Center for a New American Security published a report highlighting the national security effects of climate change.[89] These security effects include increased competition for resources between countries, mass migration from the worst affected areas, challenges to the cohesion of major states threatened by the rise in sea levels, and, as a consequence of these factors, an increased risk of armed conflict, including even nuclear conflicts.

Adaptation to global warming, Mitigation of global warming, and Kyoto Protocol The broad agreement among climate scientists that global temperatures will continue to increase has led some nations, states, corporations and individuals to implement actions to try to curtail global warming or adjust to it. Many environmental groups encourage individual action against global warming, often by the consumer, but also by community and regional organizations. Others have suggested a quota on worldwide fossil fuel production, citing a direct link between fossil fuel production and CO_2 emissions.[90][91]There has also been business action on climate change, including efforts at increased energy efficiency and limited moves towards use of alternative fuels. One important innovation has been the development of greenhouse gas emissions trading through which companies, in conjunction with government, agree to cap their emissions or to purchase credits from those below their allowances.The world's primary international agreement on combating global warming is the Kyoto Protocol, an amendment to the UNFCCC negotiated in 1997. The Protocol now covers more than 160 countries globally and over 55% of global greenhouse gas emissions.[92] Only the United States and Kazakhstan have not ratified the treaty, with the United States historically being the world's largest emitter of greenhouse gas. This treaty expires in 2012, and international talks began in May 2007 on a future treaty to succeed the current one.[93]Claiming "serious harm" to the United States economy and the exemption of "80 percent of the world, including major population centers" like China and India from the treaty, George W. Bush contends that the Kyoto Protocol is an unfair and ineffective means of addressing global climate change concerns.[94] Bush has promoted improved energy technology as a means to combat climate change,[95] and various state and city governments within the United States have begun their own initiatives to indicate support and compliance with the Kyoto Protocol on a local basis; an example of this being the Regional Greenhouse Gas Initiative.[96] China and India, though exempt from its provisions as developing countries, have ratified the Kyoto Protocol. China may have passed the U.S. in total annual greenhouse gas emissions according to some recent studies. Chinese Premier Wen Jiabao has called on the nation to redouble its efforts to tackle pollution and global warming.[97] The IPCC's Working Group III is responsible for crafting reports that deal with the mitigation of global warming and analyzing the costs and benefits of different approaches. In the 2007 IPCC Fourth Assessment Report, they conclude that no one technology or sector can be completely responsible for mitigating future warming. They find there are key practices and technologies in various sectors, such as energy supply, transportation, industry, and agriculture, that should be implemented to reduced global emissions. They estimate that stabilization of carbon dioxide equivalent between 445 and 710 ppm by 2030 will result in between a 0.6% increase and 3% decrease in global gross domestic product.[98]

Social and political debate

Global warming controversy, Politics of global warming, and Economics of global warming. Climate change denial Per capita greenhouse gas emissions in 2000, including land-use change.Per country greenhouse gas emissions in 2000, including land-use change. Over the past several years, increased awareness of the scientific findings surrounding global warming has resulted in political and economic debate.[99] Poor regions, particularly Africa, appear at greatest risk from the suggested effects of global warming, while their actual emissions have been small compared to the developed world.[100] At the same time, developing country exemptions from provisions of the Kyoto Protocol have been criticized by the United States and Australia, and used as part of a rationale for continued non-ratification by the U.S.[101] In the Western world, the idea of human influence on climate has gained wider acceptance in Europe than in the United States.[102][103]The issue of climate change has sparked debate weighing the benefits of limiting industrial emissions of greenhouse gases against the costs that such changes would entail. There has been discussion in several countries about the cost and benefits of adopting alternative energy sources in order to reduce carbon emissions.[104] Organizations and companies such as the Competitive Enterprise Institute and ExxonMobil have emphasized more conservative climate change scenarios while highlighting the potential economic cost of stricter controls.[105][106][107][108] Likewise, various environmental lobbies and a number of public figures have launched campaigns to emphasize the potential risks of climate change and promote the implementation of stricter controls. Some fossil fuel companies have scaled back their efforts in recent years,[109] or called for policies to reduce global warming.[110]Another point of contention is the degree to which newly developed economies such as India and China should be expected to constrain their emissions. China's gross national CO_2 emissions are expected to exceed those of the U.S. within the next few years, and may have already done so according to a 2006 report.[111] China has contended that it has less of an obligation to reduce emissions since its per capita emissions are roughly one-fifth that of the United States.[112] India, also exempt from Kyoto restrictions and another of the biggest sources of industrial emissions, has made similar assertions.[113] However, the U.S. contends that if they must bear the cost of reducing emissions, then China should do the same.[114]

Ocean acidification, global dimming, and ozone depletion

A variety of issues are often raised in relation to global warming. One is ocean acidification. Increased atmospheric CO_2 increases the amount of CO_2 dissolved in the oceans.[115] CO_2 dissolved in the ocean reacts with water to form carbonic acid, resulting in acidification. Ocean surface pH is estimated to have decreased from 8.25 near the beginning of the industrial era to 8.14 by 2004,[116] and is projected to decrease by a further 0.14 to 0.5 units by 2100 as the ocean absorbs more CO_2.[1][117] Since organisms and ecosystems are adapted to a narrow range of pH, this raises extinction concerns, directly driven by increased atmospheric CO_2, that could disrupt food webs and impact human societies that depend on marine ecosystem services.[118]Global dimming, the gradual reduction in the amount of global direct irradiance at the Earth's surface, may have partially mitigated global warming in the late twentieth century. From 1960 to 1990 human-caused aerosols likely precipitated this effect. Scientists have stated with 66–90% confidence that the effects of human-caused aerosols, along with volcanic activity, have offset some of the global warming, and that greenhouse gases would have resulted in more warming than observed if not for these dimming agents.[1]Ozone depletion, the steady decline in the total amount of ozone in Earth's stratosphere, is frequently cited in relation to global warming. Although there are areas of linkage, the relationship between the two is not strong.

NOTES AND REFERENCES

1. Summary for Policymakers (PDF). *Climate Change 2007: The Physical Science Basis. Contribution of Working Group I to the Fourth Assessment Report of the Intergovernmental Panel on Climate Change.* Intergovernmental Panel on Climate Change (2007-02-05). Retrieved on 2007-02-02. "The updated hundred-year linear trend (1906 to 2005) of 0.74 °C [0.56 °C to 0.92 °C] is therefore larger than the corresponding trend for 1901 to 2000 given in the TAR of 0.6 °C [0.4 °C to 0.8 °C]."

2. Hegerl, Gabriele C.; *et al.* (2007-05-07). *Understanding and Attributing Climate Change* (PDF). *Climate Change 2007: The Physical Science Basis. Contribution of Working Group I to the Fourth Assessment Report of the Intergovernmental Panel on Climate Change* 690. Intergovernmental Panel on Climate Change. Retrieved on 2007-05-20. "Recent estimates (Figure 9.9) indicate a

relatively small combined effect of natural forcings on the global mean temperature evolution of the seconds half of the twentieth century, with a small net cooling from the combined effects of solar and volcanic forcings"

3. Ammann, Caspar; *et al.* (2007-04-06). "Solar influence on climate during the past millennium: Results from ransient simulations with the NCAR Climate Simulation Model" (PDF). *Proceedings of the National Academy of Sciences of the United States of America* 104 (10): 3713–3718. "However, because of a lack of interactive ozone, the model cannot fully simulate features discussed in (44)." "While the NH temperatures of the high-scaled experiment are often colder than the lower bound from proxy data, the modeled decadal-scale NH surface temperature for the medium-scaled case falls within the uncertainty range of the available temperature reconstructions. The medium-scaled simulation also broadly reproduces the main features seen in the proxy records." "Without anthropogenic forcing, the 20th century warming is small. The simulations with only natural forcing components included yield an early 20th century peak warming of H"0.2 °C (H"1950 AD), which is reduced to about half by the end of the century because of increased volcanism."

4. The 2001 joint statement was signed by the scientific academies of Australia, Belgium, Brazil, Canada, the Caribbean, China, France, Germany, India, Indonesia, Ireland, Italy, Malaysia, New Zealand, Sweden, and the UK. The 2005 statement added Japan, Russia, and the U.S. The 2007 statement added Mexico, and South Africa. Professional societies include American Meteorological Society, American Geophysical Union, American Institute of Physics, American Astronomical Society, American Association for the Advancement of Science, Stratigraphy Commission of the Geological Society of London, Geological Society of America, American Chemical Society, and Engineers Australia.

5. The Science Of Climate Change. Royal Society (May 2001). Retrieved on 2008-01-04.

6. Joint science academies' statement: Global response to

climate change. Royal Society (June 2005). Retrieved on 2008-01-04.

7. Joint science academies' statement on growth and responsibility: sustainability, energy efficiency and climate protection. Potsdam Institute for Climate Impact Research (May 2007). Retrieved on 2008-01-04.

8. Don't fight, adapt. National Post (December 2007). Retrieved on 2007-11-18.

9. A guide to facts and fictions about climate change. Royal Society (March 2005). Retrieved on 2007-11-18. ""However, the overwhelming majority of scientists who work on climate change agree on the main points""

10. Beyond the Ivory Tower: The Scientific Consensus on Climate Change. Science Magazine (December 2004). Retrieved on 2008-01-04.

11. http://www.merriam-webster.com/dictionary/global+warming

12. Climate Change: Basic Information. United States Environmental Protection Agency (2006-12-14). Retrieved on 2007-02-09. "In common usage, 'global warming' often refers to the warming that can occur as a result of increased emissions of greenhouse gases from human activities."

13. United Nations Framework Convention on Climate Change, Article I. United Nations Framework Convention on Climate Change. Retrieved on 2007-01-15.

14. Berger, A.; *et al.* (2005-12-10). "On the origin of the 100-kyr cycles in the astronomical forcing". *Paleoceanography* 20 (4). PA4019. Retrieved on 2007-11-05.

15. Genthon, C.; *et al.* (1987-10-01). "Vostok Ice Core - Climatic response to CO_2 and orbital forcing changes over the last climatic cycle". *Nature* 329 (6138): 414–418. Retrieved on 2007-11-05.

16. Alley, Richard B.; *et al.* (January 2002). "A northern lead in the orbital band: north-south phasing of Ice-Age events". *Quaternary Science Reviews* 21 (1-3): 431–441. Retrieved

on 2007-11-05.

17. Robock, Alan, and Clive Oppenheimer, Eds., 2003: Volcanism and the Earth's Atmosphere, Geophysical Monograph 139, American Geophysical Union, Washington, DC, 360 pp.

18. Joint science academies' statement: The science of climate change (ASP). Royal Society (2001-05-17). Retrieved on 2007-04-01. "The work of the Intergovernmental Panel on Climate Change (IPCC) represents the consensus of the international scientific community on climate change science"

19. "Rising to the climate challenge" (2007-10-18). *Nature* 449 (7164). Retrieved on 2007-11-06.

20. Svensmark, Henrik (February 2007). "Cosmoclimatology: a new theory emerges". *Astronomy & Geophysics* 48 (1): 18-24. doi:10.1111/j.1468-4004.2007.48118.x.

21. Forster, Piers; *et al.* (2007-02-05). Changes in Atmospheric Constituents and in Radiative Forcing (PDF). *Climate Change 2007: The Physical Science Basis. Contribution of Working Group I to the Fourth Assessment Report of the Intergovernmental Panel on Climate Change* 188-193. Intergovernmental Panel on Climate Change. Retrieved on 2007-09-17.

22. Bard, Edouard; Frank, Martin (2006-06-09). "Climate change and solar variability: What's new under the sun?". *Earth and Planetary Science Letters* **248** (1-2): 1-14. Retrieved on 2007-09-17.

23. Meehl, Gerald A.; *et al.* (2005-03-18). "How Much More Global Warming and Sea Level Rise" (PDF). *Science* **307** (5716): 1769–1772. doi:10.1126/science.1106663. Retrieved on 2007-02-11.

24. IPCC WG1 AR4 Report — Chapter 1: Historical Overview of Climate Change Science (PDF). *IPCC WG1 AR4 Report* p97 (pdf page 5 of 36). IPCC (2007). Retrieved on 2007-10-07. "To emit 240 W m–2, a surface would have to have a temperature of around –19 °C. This is much colder than the conditions that actually exist at the Earth's surface

(the global mean surface temperature is about 14 °C). Instead, the necessary –19 °C is found at an altitude about 5 km above the surface."

25. Note that the Greenhouse Effect produces a temperature increase of about 33 °C (59 °F) with respect to black body predictions and not a surface temperature of 33 °C (91 °F) which is 32 °F (0 °C) higher. The average surface temperature is about 14 °C (57 °F). Also note that both the Celsius and Fahrenheit temperatures are expressed to 2 significant figures even though the conversion formula produces 3.

26. Kiehl, J. T.; Kevin E. Trenberth (February 1997). "Earth's Annual Global Mean Energy Budget" (PDF). *Bulletin of the American Meteorological Society* 78 (2): 197-208. Retrieved on 2006-05-01.

27. Water vapour: feedback or forcing?. RealClimate (6 Apr 2005). Retrieved on 2006-05-01.

28. Neftel, A., E. Moor, H. Oeschger, and B. Stauffer. (1985). "Evidence from polar ice cores for the increase in atmospheric CO2 in the past two centuries". *Nature* 315:45-47.

29. Pearson, Paul N.; Palmer, Martin R. (2000-08-17). "Atmospheric carbon dioxide concentrations over the past 60 million years". *Nature* 406 (6797): 695–699.

30. Summary for Policymakers. *Climate Change 2001: The Scientific Basis. Contribution of Working Group I to the Third Assessment Report of the Intergovernmental Panel on Climate Change.* Intergovernmental Panel on Climate Change (2001-01-20). Retrieved on 2007-01-18.

31. Tans, Pieter. Trends in Atmospheric Carbon Dioxide – Mauna Loa. National Oceanic and Atmospheric Administration. Retrieved on 2008-02-15.

32. Prentice, I. Colin; *et al.* (2001-01-20). 3.7.3.3 SRES scenarios and their implications for future CO2 concentration. *Climate Change 2001: The Scientific Basis. Contribution of Working Group I to the Third Assessment Report of the Intergovernmental Panel on Climate Change.*

Intergovernmental Panel on Climate Change. Retrieved on 2007-04-28.

33. 4.4.6. Resource Availability. *IPCC Special Report on Emissions Scenarios*. Intergovernmental Panel on Climate Change. Retrieved on 2007-04-28.

34. Soden, Brian J.; Held, Isacc M. (2005-11-01). "An Assessment of Climate Feedbacks in Coupled Ocean–Atmosphere Models" (PDF). *Journal of Climate* 19 (14). Retrieved on 2007-04-21. "Interestingly, the true feedback is consistently weaker than the constant relative humidity value, implying a small but robust reduction in relative humidity in all models on average" "clouds appear to provide a positive feedback in all models"

35. Panel on Climate Change Feedbacks, Climate Research Committee, National Research Council, 2004: Understanding Climate Change Feedbacks.

36. Stocker, Thomas F.; *et al.* (2001-01-20). 7.5.2 Sea Ice. *Climate Change 2001: The Scientific Basis. Contribution of Working Group I to the Third Assessment Report of the Intergovernmental Panel on Climate Change*. Intergovernmental Panel on Climate Change. Retrieved on 2007-02-11.

37. Sample, Ian. "Warming Hits 'Tipping Point'", The Guardian, 2005-08-11. Retrieved on 2007-01-18.

38. Buesseler, Ken O.; *et al.* (2007-04-27). "Revisiting Carbon Flux Through the Ocean's Twilight Zone". *Science* 316 (5824): 567–570. Retrieved on 2007-11-16.

39. Scafetta, Nicola; West, Bruce J. (2006-03-09). "Phenomenological solar contribution to the 1900–2000 global surface warming" (PDF). *Geophysical Research Letters* 33 (5). doi:10.1029/2005GL025539. L05708. Retrieved on 2007-05-08.

40. Stott, Peter A.; *et al.* (2003-12-03). "Do Models Underestimate the Solar Contribution to Recent Climate Change?". *Journal of Climate* 16 (24): 4079–4093. doi:10.1175/1520-

0442(2003)016%3C4079: DMUTSC%3E2.0.CO;2. Retrieved on 2007-04-16.

41. Marsh, Nigel; Henrik, Svensmark (November 2000). "Cosmic Rays, Clouds, and Climate" (PDF). *Space Science Reviews* 94: 215–230. doi:10.1023/A:1026723423896. Retrieved on 2007-04-17.

42. Svensmark, Henrik (July 2000). "Cosmic Rays and Earth's Climate" (PDF). *Space Science Reviews* **93** (1-2): 175-185. Retrieved on 2007-09-17.

43. Hegerl, Gabriele C.; *et al.*. Understanding and Attributing Climate Change (PDF). *Climate Change 2007: The Physical Science Basis. Contribution of Working Group I to the Fourth Assessment Report of the Intergovernmental Panel on Climate Change* 675. Intergovernmental Panel on Climate Change. Retrieved on 2008-02-01.

44. Climate Change 2001:Working Group I: The Scientific Basis (Fig. 2.12) (2001). Retrieved on 2007-05-08.

45. Ozone History

46. Foukal, Peter; *et al.* (2006-09-14). "Variations in solar luminosity and their effect on the Earth's climate.". *Nature*. Retrieved on 2007-04-16.

47. National Center for Atmospheric Research (2006-09-14). "Changes in Solar Brightness Too Weak to Explain Global Warming". Press release. Retrieved on 2007-07-13.

48. Lockwood, Mike; Claus Fröhlich. "Recent oppositely directed trends in solar climate forcings and the global mean surface air temperature". *Proceedings of the Royal Society A*. doi:10.1098/rspa.2007.1880. Retrieved on 2007-07-21. "Our results show that the observed rapid rise in global mean temperatures seen after 1985 cannot be ascribed to solar variability, whichever of the mechanisms is invoked and no matter how much the solar variation is amplified."

49. Reply to Lockwood and Fröhlich - The persistent role of the Sun in climate forcing — Spacecenter

50. Richard Black. "'No Sun link' to climate change", BBC News Online, 3 April 2008.

51. T Sloan and A W Wolfendale. "Testing the proposed causal link between cosmic rays and cloud cover". *Environ. Res. Lett.* 3. doi:10.1088/1748-9326/3/2/024001.

52. preprint of this paper can be found here

53. Working group I, section 3.2.2.2 of the 2007 IPPC page 243

54. Smith, Thomas M.; Reynolds, Richard W. (2005-05-15). "A Global Merged Land–Air–Sea Surface Temperature Reconstruction Based on Historical Observations (1880–1997)" (PDF). *Journal of Climate* 18 (12): 2021–2036. ISSN 0894-8755. Retrieved on 2007-03-14.

55. Rowan T. Sutton, Buwen Dong, Jonathan M. Gregory (2007). "Land/sea warming ratio in response to climate change: IPCC AR4 model results and comparison with observations". *Geophysical Research Letters* 34. doi:10.1029/2006GL028164. Retrieved on 2007-09-19.

56. Hansen, James E.; *et al.* (2006-01-12). Goddard Institute for Space Studies, GISS Surface Temperature Analysis. NASA Goddard Institute for Space Studies. Retrieved on 2007-01-17.

57. Global Temperature for 2005: second warmest year on record (PDF). Climatic Research Unit, School of Environmental Sciences, University of East Anglia (2005-12-15). Retrieved on 2007-04-13.

58. WMO STATEMENT ON THE STATUS OF THE GLOBAL CLIMATE IN 2005 (PDF). World Meteorological Organization (2005-12-15). Retrieved on 2007-04-13.

59. Data @ NASA GISS: GISS Surface Temperature Analysis: 2005 Summation

60. Mitchell, J. F. B.; *et al.* (2001-01-20). 12.4.3.3 Space-time studies. *Climate Change 2001: The Scientific Basis. Contribution of Working Group I to the Third Assessment Report of the Intergovernmental Panel on Climate Change.*

Intergovernmental Panel on Climate Change. Retrieved on 2007-01-04.

61. Global warming in the twenty-first century: An alternative scenario - Hansen et al. 97 (18): 9875 - Proceedings of the National Academy of Sciences

62. Ruddiman, William F. (March 2005). "How Did Humans First Alter Global Climate?" (PDF). *Scientific American* 292 (3): 46–53. Retrieved on 2007-03-05.

63. Schmidt, Gavin; *et al.* (2004-12-10). "A note on the relationship between ice core methane concentrations and insolation". *Geophysical Research Letters* 31 (23). doi:10.1029/2004GL021083. L23206. Retrieved on 2007-03-05.

64. Hansen, James; *et al.* (2006-09-26). "Global temperature change" (PDF). *PNAS* 103: 14288–14293. Retrieved on 2007-04-20.

65. Open University (2004-01-30). "The Open University Provides Answers on Global Warming" (PDF). Press release. Retrieved on 2007-03-04.

66. Cohen, Anthony S.; *et al.* (February 2004). "Osmium isotope evidence for the regulation of atmospheric CO_2 by continental weathering" (PDF). *Geology* **32** (2): 157–160. doi:10.1130/G20158.1. Retrieved on 2007-03-04.

67. Chapter 7, "Couplings Between Changes in the Climate System and Biogeochemistry" (PDF). *Fourth Assessment Report of the Intergovernmental Panel on Climate Change.* Intergovernmental Panel on Climate Change (2007-02-05). Retrieved on 2008-02-21.

68. Hansen, James (2000). Climatic Change: Understanding Global Warming. *One World: The Health & Survival of the Human Species in the 21st Century.* Health Press. Retrieved on 2007-08-18.

69. Summary for Policymakers. *Climate Change 2001: The Scientific Basis. Contribution of Working Group I to the Third Assessment Report of the Intergovernmental Panel*

on Climate Change. Intergovernmental Panel on Climate Change (2001-01-20). Retrieved on 2007-04-28.

70. Torn, Margaret; Harte, John (2006-05-26). "Missing feedbacks, asymmetric uncertainties, and the underestimation of future warming". *Geophysical Research Letters* **33** (10). L10703. Retrieved on 2007-03-04.

71. Harte, John; *et al.* (2006-10-30). "Shifts in plant dominance control carbon-cycle responses to experimental warming and widespread drought". *Environmental Research Letters* **1** (1). 014001. Retrieved on 2007-05-02.

72. Scheffer, Marten; *et al.* (2006-05-26). "Positive feedback between global warming and atmospheric CO2 concentration inferred from past climate change.". Geophysical Research Letters 33. doi:10.1029/2005gl025044. Retrieved on 2007-05-04.

73. Stocker, Thomas F.; *et al.* (2001-01-20). 7.2.2 Cloud Processes and Feedbacks. *Climate Change 2001: The Scientific Basis. Contribution of Working Group I to the Third Assessment Report of the Intergovernmental Panel on Climate Change*. Intergovernmental Panel on Climate Change. Retrieved on 2007-03-04.

74. Macey, Jennifer. "Global warming opens up Northwest Passage", ABC News, September 19, 2007. Retrieved on 2007-12-11.

75. *Climate Change 2001: Impacts, Adaptation and Vulnerability. Contribution of Working Group II to the Third Assessment Report of the Intergovernmental Panel on Climate Change*. Intergovernmental Panel on Climate Change (2001-02-16). Retrieved on 2007-03-14.

76. McMichael AJ, Woodruff RE, Hales S (2006). "Climate change and human health: present and future risks". *Lancet* 367 (9513): 859-69. doi:10.1016/S0140-6736(06)68079-3. PMID 16530580.

77. *Summary for Policymakers* (PDF). *Climate Change 2007: Impacts, Adaptation and Vulnerability. Working Group II*

Contribution to the Intergovernmental Panel on Climate Change Fourth Assessment Report. Intergovernmental Panel on Climate Change (2007-04-13). Retrieved on 2007-04-28.

78. Church, John A.; *et al.* (2001-01-20). Executive Summary of Chapter 11. *Climate Change 2001: The Scientific Basis. Contribution of Working Group I to the Third Assessment Report of the Intergovernmental Panel on Climate Change*. Intergovernmental Panel on Climate Change. Retrieved on 2005-12-19.

79. Thomas, Chris D.; *et al.* (2004-01-08). "Extinction risk from climate change" (PDF). *Nature* 427 (6970): 145-138. doi:10.1038/nature02121. Retrieved on 2007-03-18.

80. McLaughlin, John F.; *et al.* (2002-04-30). "Climate change hastens population extinctions" (PDF). *PNAS* 99 (9): 6070–6074. doi:10.1073/pnas.052131199. Retrieved on 2007-03-29.

81. Botkin, Daniel B.; *et al.* (March 2007). "Forecasting the Effects of Global Warming on Biodiversity". *BioScience* **57** (3): 227–236. doi:10.1641/B570306. Retrieved on 2007-11-30.

82. At-a-glance: The Stern Review. BBC (2006-10-30). Retrieved on 2007-04-29.

83. Tol and Yohe (2006) "A Review of the Stern Review" *World Economics* **7**(4): 233-50. See also other critiques in *World Economics* **7**(4).

84. J. Bradford DeLong. Do unto others....

85. John Quiggin. Stern and the critics on discounting.

86. Terry Barker (April 14, 2008). Full quote from IPCC on costs of climate change. FT.com. Retrieved on 2008-04-14.

87. Dlugolecki, Andrew; *et al.* (2002). Climate Risk to Global Economy (PDF). *CEO Briefing: UNEP FI Climate Change Working Group*. United Nations Environment Programme. Retrieved on 2007-04-29.

88. Thomas Schelling: Developing Countries Will Suffer Most from Global Warming. *Resources 164*. Retrieved on 2008-03-01.

89. Kurt M. Campbell, Jay Gulledge, J.R. McNeill, John Podesta, Peter Ogden, Leon Fuerth, R. James Woolsey, Alexander T.J. Lennon, Julianne Smith, Richard Weitz, Derek Mix (Oktober 2007). The Age of Consequences: The Foreign Policy and National Security Implications of Global Climate Change (PDF). Retrieved on 2007-11-06.

90. Climate Control: a proposal for controlling global greenhouse gas emissions (PDF). Sustento Institute. Retrieved on 2007-12-10.

91. Monbiot, George. Rigged - The climate talks are a stitch-up, as no one is talking about supply. (HTML). Retrieved on 2007-12-22.

92. Kyoto Protocol Status of Ratification (PDF). United Nations Framework Convention on Climate Change (2006-07-10). Retrieved on 2007-04-27.

93. Climate talks face international hurdles, by Arthur Max, Associated press, 5/14/07.

94. George W. Bush (March 13, 2001). Text of a Letter from the President to Senators Hagel, Helms, Craig, and Roberts. Office of the Press Secretary. Retrieved on 2007-11-21.

95. State of the Union Address, retrieved 2008-01-28. "The United States is committed to strengthening our energy security and confronting global climate change. And the best way to meet these goals is for America to continue leading the way toward the development of cleaner and more energy-efficient technology."

96. Regional Greenhouse Gas Initiative. Retrieved on 2006-11-07.

97. Wen Urges Greater China Effort to Fight Pollution, July 10, 2007, Reuters, via planetark.com. Retrieved on 2007-07-27.

98. Summary for Policymakers (PDF). *Climate Change 2007: The Physical Science Basis. Contribution of Working Group III to the Fourth Assessment Report of the Intergovernmental Panel on Climate Change.* Intergovernmental Panel on Climate Change (2007-05-04). Retrieved on 2007-12-09.

99. Weart, Spencer (2006), "The Public and Climate Change", in Weart, Spencer, The Discovery of Global Warming, American Institute of Physics, <http://www.aip.org/history/climate/Public.htm>. Retrieved on 14 April 2007

100. Revkin, Andrew. "Poor Nations to Bear Brunt as World Warms", The New York Times, 2007-04-01. Retrieved on 2007-05-02.

101. Brahic, Catherine (2006-04-25). China's emissions may surpass the US in 2007. New Scientist. Retrieved on 2007-05-02.

102. Crampton, Thomas. "More in Europe worry about climate than in U.S., poll shows", International Herald Tribune, 2007-01-04. Retrieved on 2007-04-14.

103. Summary of Findings. *Little Consensus on Global Warming. Partisanship Drives Opinion.* Pew Research Center (2006-07-12). Retrieved on 2007-04-14.

104. "EU agrees on carbon dioxide cuts", BBC, 2007-03-09. Retrieved on 2007-05-04.

105. Begley, Sharon. "The Truth About Denial", Newsweek, 2007-08-13. Retrieved on 2007-08-13.

106. Adams, David (2006-09-20). Royal Society tells Exxon: stop funding climate change denial. The Guardian. Retrieved on 2007-08-09.

107. "Exxon cuts ties to global warming skeptics", MSNBC, 2007-01-12. Retrieved on 2007-05-02.

108. Sandell, Clayton. "Report: Big Money Confusing Public on Global Warming", ABC, 2007-01-03. Retrieved on 2007-04-27.

109. "Greenpeace: Exxon still funding climate skeptics", USA

Today, 2007-05-18. Retrieved on 2007-07-09.

110. Ceres (April 28, 2004). "Global Warming Resolutions at U.S. Oil Companies Bring Policy Commitments from Leaders, and Record High Votes at Laggards". Press release. Retrieved on 2007-07-27.

111. "Group: China tops world in CO2 emissions", Associated Press, 2007-06-20. Retrieved on 2007-10-16. ; "Group: China surpassed US in carbon emissions in 2006: Dutch report", Reuters, 2007-06-20. Retrieved on 2007-10-16.

112. China: US should take lead on climate, by Michael Casey, Associated Press, via newsvine.com 12/7/07.

113. India's glaciers give grim message on warming, by Somni Sengupta, 7/17/07, New York Times via oregonlive.com.

114. Chinese object to climate draft, BBC, 5/1/07; In Battle for U.S. Carbon Caps, Eyes and Efforts Focus on China,by Steven Mufson, Washington Post, 6/6/07.

115. The Ocean and the Carbon Cycle. *NASA* (2005-06-21). Retrieved on 2007-03-04.

116. Jacobson, Mark Z. (2005-04-02). "Studying ocean acidification with conservative, stable numerical schemes for nonequilibrium air-ocean exchange and ocean equilibrium chemistry" (PDF). *Journal of Geophysical Research* (D7). doi:10.1029/2004JD005220. D07302. Retrieved on 2007-04-28.

117. Caldeira, Ken; Wickett, Michael E. (2005-09-21). "Ocean model predictions of chemistry changes from carbon dioxide emissions to the atmosphere and ocean". *Journal of Geophysical Research* 110 (C09S04): 1–12. Retrieved on 2006-02-14.

118. Raven, John A.; *et al.* (2005-06-30). "Ocean acidification due to increasing atmospheric carbon dioxide" (ASP). Royal Society. Retrieved on 2007-05-04.

5

Causes of Global Warming

Global Warming Global warming is an increase in the average temperature of Earth's surface. Since the late 1800's, the global average temperature has increased about 0.7 to 1.4 degrees F (0.4 to 0.8 degrees C). Many experts estimate that the average temperature will rise an additional 2.5 to 10.4 degrees F (1.4 to 5.8 degrees C) by 2100. That rate of increase would be much larger than most past rates of increase.

Scientists worry that human societies and natural ecosystems might not adapt to rapid climate changes. An ecosystem consists of the living organisms and physical environment in a particular area. Global warming could cause much harm, so countries throughout the world drafted an agreement called the Kyoto Protocol to help limit it.

Causes of global warming

Climatologists (scientists who study climate) have analyzed the global warming that has occurred since the late 1800's. A majority of climatologists have concluded that human activities are responsible for most of the warming. Human activities contribute to global warming by enhancing Earth's natural greenhouse effect. The greenhouse effect warms Earth's surface through a complex process involving sunlight, gases, and particles in the atmosphere. Gases that trap heat in the atmosphere are known as greenhouse gases.

The main human activities that contribute to global warming are the burning of fossil fuels (coal, oil, and natural gas) and the clearing of land. Most of the burning occurs in automobiles, in factories, and in electric power plants that provide energy for houses and office buildings. The burning of fossil fuels creates carbon dioxide, whose chemical formula is CO_2. CO_2 is a greenhouse gas that slows the escape of heat into space. Trees and other plants remove CO_2 from the air during photosynthesis, the process they use to produce food. The clearing of land contributes to the buildup of CO_2 by reducing the rate at which the gas is removed from the atmosphere or by the decomposition of dead vegetation.

A small number of scientists argue that the increase in greenhouse gases has not made a measurable difference in the temperature. They say that natural processes could have caused global warming. Those processes include increases in the energy emitted (given off) by the sun. But the vast majority of climatologists believe that increases in the sun's energy have contributed only slightly to recent warming.

The impact of global warming

Thousands of icebergs float off the coast of the Antarctic Peninsula after 1,250 square miles (3,240 square kilometers) of the Larsen B ice shelf disintegrated in 2002. The area of the ice was larger than the state of Rhode Island or the nation of Luxembourg. Antarctic ice shelves have been shrinking since the early 1970's because of climate warming in the region. Image credit: NASA/Earth Observatory

Continued global warming could have many damaging effects. It might harm plants and animals that live in the sea. It could also force animals and plants on land to move to new habitats. Weather patterns could change, causing flooding, drought, and an increase in damaging storms. Global warming could melt enough polar ice to raise the sea level. In certain parts of the world, human disease could spread, and crop yields could decline.

Harm to ocean life

Through global warming, the surface waters of the oceans could become warmer, increasing the stress on ocean ecosystems, such as coral reefs. High water temperatures can cause a damaging process called coral bleaching. When corals bleach, they expel the algae that give them their color and nourishment. The corals turn white and, unless the water temperature cools, they die. Added warmth also helps spread diseases that affect sea creatures.

Changes of habitat

Widespread shifts might occur in the natural habitats of animals and plants. Many species would have difficulty surviving in the regions they now inhabit. For example, many flowering plants will not bloom without a sufficient period of winter cold. And human occupation has altered the landscape in ways that would make new habitats hard to reach or unavailable altogether.

Weather damage

Extreme weather conditions might become more frequent and therefore more damaging. Changes in rainfall patterns could increase both flooding and drought in some areas. More hurricanes and other tropical storms might occur, and they could become more powerful.

Rising sea level

Continued global warming might, over centuries, melt

large amounts of ice from a vast sheet that covers most of West Antarctica. As a result, the sea level would rise throughout the world. Many coastal areas would experience flooding, erosion, a loss of wetlands, and an entry of seawater into freshwater areas. High sea levels would submerge some coastal cities, small island nations, and other inhabited regions.

Threats to human health

Tropical diseases, such as malaria and dengue, might spread to larger regions. Longer-lasting and more intense heat waves could cause more deaths and illnesses. Floods and droughts could increase hunger and malnutrition.

Changes in crop yields

Canada and parts of Russia might benefit from an increase in crop yields. But any increases in yields could be more than offset by decreases caused by drought and higher temperatures — particularly if the amount of warming were more than a few degrees Celsius. Yields in the tropics might fall disastrously because temperatures there are already almost as high as many crop plants can tolerate.

Limited global warming

Climatologists are studying ways to limit global warming. Two key methods would be (1) limiting CO2 emissions and (2) carbon sequestration — either preventing carbon dioxide from entering the atmosphere or removing CO2 already there.

Limiting CO2 emissions

Two effective techniques for limiting CO2 emissions would be (1) to replace fossil fuels with energy sources that do not emit CO2, and (2) to use fossil fuels more efficiently.

Alternative energy sources that do not emit CO2 include the wind, sunlight, nuclear energy, and underground steam. Devices known as wind turbines can convert wind energy to electric energy. Solar cells can convert sunlight to electric energy, and various devices can convert solar energy to useful heat. Geothermal power plants convert energy in underground steam to electric energy.

Alternative sources of energy are more expensive to use than fossil fuels. However, increased research into their use would almost certainly reduce their cost.

Carbon sequestration could take two forms: (1) underground or underwater storage and (2) storage in living plants.

Underground or underwater storage would involve injecting industrial emissions of CO2 into underground geologic formations or the ocean. Suitable underground formations include natural reservoirs of oil and gas from which most of the oil or gas has been removed. Pumping CO2 into a reservoir would have the added benefit of making it easier to remove the remaining oil or gas. The value of that product could offset the cost of sequestration. Deep deposits of salt or coal could also be suitable.

The oceans could store much CO2. However, scientists have not yet determined the environmental impacts of using the ocean for carbon sequestration.

Storage in living plants

Green plants absorb CO2 from the atmosphere as they grow. They combine carbon from CO2 with hydrogen to make simple sugars, which they store in their tissues. After plants die, their bodies decay and release CO2. Ecosystems with abundant plant life, such as forests and even cropland, could tie up much carbon. However, future generations of people would have to keep the ecosystems intact. Otherwise, the sequestered carbon would re-enter the atmosphere as CO2.

Agreement on global warming

Delegates from more than 160 countries met in Kyoto, Japan, in 1997 to draft the agreement that became known as the Kyoto Protocol. That agreement calls for decreases in the emissions of greenhouse gases.

Emissions targets

Thirty-eight industrialized nations would have to restrict their emissions of CO2 and five other greenhouse gases. The restrictions would occur from 2008 through 2012. Different countries would have different emissions targets. As a whole, the 38 countries would restrict their emissions to a yearly average of about 95 percent of their 1990 emissions. The agreement does not place restrictions on developing countries. But it encourages the industrialized nations to cooperate in helping developing countries limit emissions voluntarily.

Industrialized nations could also buy or sell emission reduction units. Suppose an industrialized nation cut its emissions more than was required by the agreement. That country could sell other industrialized nations emission reduction units allowing those nations to emit the amount equal to the excess it had cut.

Several other programs could also help an industrialized nation earn credit toward its target. For example, the nation might help a developing country reduce emissions by replacing fossil fuels in some applications.

Approving the agreement

The protocol would take effect as a treaty if (1) at least 55 countries ratified (formally approved) it, and (2) the industrialized countries ratifying the protocol had CO2 emissions in 1990 that equaled at least 55 percent of the emissions of all 38 industrialized countries in 1990.

In 2001, the United States rejected the Kyoto Protocol. President George W. Bush said that the agreement could harm the U.S. economy. But he declared that the United States would work with other countries to limit global warming. Other countries, most notably the members of the European Union, agreed to continue with the agreement without United States participation.

By 2004, more than 100 countries, including nearly all the countries classified as industrialized under the protocol, had ratified the agreement. However, the agreement required ratification by Russia or the United States to go into effect. Russia ratified the protocol in November 2004. The treaty was to come into force in February 2005.

Analyzing global warming

Scientists use information from several sources to analyze global warming that occurred before people began to use thermometers. Those sources include tree rings, cores (cylindrical samples) of ice drilled from Antarctica and Greenland, and cores drilled out of sediments in oceans. Information from these sources indicates that the temperature increase of the 1900's was probably the largest in the last 1,000 years.

Computers help climatologists analyze past climate changes and predict future changes. First, a scientist programs a computer with a set of mathematical equations known as a climate model. The equations describe how various factors, such as the amount of CO2 in the atmosphere, affect the temperature of Earth's surface. Next, the scientist enters data representing the values of those factors at a certain time. He or she then runs the program, and the computer describes how the temperature would vary. A computer's representation of changing climatic conditions is known as a climate simulation.

In 2001, the Intergovernmental Panel on Climate Change (IPCC), a group sponsored by the United Nations (UN), published results of climate simulations in a report on global warming. Climatologists used three simulations to determine whether natural variations in climate produced the warming of the past 100 years. The first simulation took into account both natural processes and human activities that affect the climate. The second simulation took into account only the natural processes, and the third only the human activities.

The climatologists then compared the temperatures predicted by the three simulations with the actual temperatures recorded by thermometers. Only the first simulation, which took into account both natural processes and human activities, produced results that corresponded closely to the recorded temperatures.

The IPCC also published results of simulations that predicted temperatures until 2100. The different simulations took into account the same natural processes but different patterns of human activity. For example, scenarios differed in the amounts of CO2 that would enter the atmosphere due to human activities.

The simulations showed that there can be no "quick fix" to the problem of global warming. Even if all emissions of greenhouse gases were to cease immediately, the temperature would continue to increase after 2100 because of the greenhouse gases already in the atmosphere.

Strategy and Response of India

India's foreign environmental policy has been mainly shaped in large measures by two main sources. First, Orthodoxy, established by Mrs Indira Gandhi, regarding the way in which environmental problems in poor countries need to viewed, and the legacy of India's foreign policy.

Prime Minister Indira Gandhi's addressed to the 1972 UN conference on Human Environment in Stockholm established, an orthodoxy that has prevailed ever since in India's foreign environmental policy. International support also helped to sustain the orthodoxy that emerged from Mrs. Gandhi's views. These views reflected, and were reinforced by Third World views on environment and development. Mrs. Gandhi's views also gained support from liberal elements in the North who accepted as legitimate, priority that developing countries accorded to development. Thus, the wide publicity given to Mrs. Gandhi's Stockholm speech reinforced the Indian government's espousal of the orthodoxy that emerged from it.

The traditions of foreign policy have had a major influence on India's foreign environmental policy, because of the Congress Party rule for about forty-five years since independence in 1947 and the leadership of the congress patty by three generations of the Nehru -Gandhi family. Foreign Policy traditions have also been maintained because they are believed to have served the country well, and have enjoyed strong cross party Support. This is illustrated by the lack of change in foreign policy despite changes in governments. Finally, the MEA's lack of environmental expertise and more generally, its defencies in terms of institutionalised forward thinking and policy planning, have made it particularly prone to fall back on tradition as a guide to policy. Indeed, many observers agree that few of India's major actions in the field of foreign affairs since 1971 have resulted from long-term strategic planning.

In the policy making process of India there is extremely limited impact of pluralist inputs. Domestic non-state actors such as the media, NGOs and business interests generally had very little influence over policy formation. Therefore the government is left with considerable degree of autonomy in policy making.

India's delegations to the international negotiations on climate change are normally composed of middle and top level officials from the MOEF. These officials accompanied by a numbers of other representatives from India's Permanent Mission to the United Nations, and in addition there are often one or two representatives from the Ministry of External Affairs' United Nations Division. The UN Division of the MEA is one of the nine functional divisions. It deals with all aspects of India's relation with the UN. The Division is an intermediary agency between India's permanent Mission of UN and the political leadership whom it helps in arriving at decisions. However, during the UNCED preparations, MEA's senior officials led delegations to the INC. But with the departure of this official from the ministry, the MEA lost its leadership and the MOEF took over. At present, India's delegations to climate change negotiations are usually led by the minister of the Environment or a top level official from MOEF.

Prior to UNCED the MOEF and MEA had a dispute over who was to head India's delegations and be in charge of shaping the national position. The MEA had the upper hand during the INC negotiations; but since Rio the MOEF has been the ministry in charge of the Indian national position. The MEA's task is to ensure that, the MOEF position on climate change does not take the Indian government beyond its general line of foreign policy.

The MOEF receives technical support from a wide range of ministries i.e., Ministry of Power, the Ministry of Coal and Ministry of Non-Conventional Energy Resource, and the technical inputs provided by Ministry of Science and Technology and Ministry and Finance. The participation of ministry simply fulfils the functions of providing information relating to technical matters. But the actual policy making process lies with MOEF, only.

During the initial policy making on climate change the MOEF leaned heavily on the advice of a number of environmental organisation and climate change experts. The MOEF had little knowledge of the international or domestic issues at stake. NGO's had very limited influence, because very few NGOs which posses at least some of the resources and the interests made significant contribution to Indian policy.

In the climate change issues, the CSE and TERI appeared to have made substantial contributions. The CSE's robust criticism of the WRI's 1990 report encouraged the government to reject the WRI's statistics and demand the application of the principle of per capita equity in green house gas allocations. The CSE however highlighted problems in the methodology adopted by the WRI. It criticised the WRI for focusing on current annual emissions, rather than cumulative emissions; by neglecting the past, the WRI exaggerated the South's responsibility for climate change. The CSE also pointed out that by focusing only on their short term heating effect, the WRI was ignoring the various atmospheric residence times and long term warming effect of greenhouse emissions; consequently, the WRI report was biased against countries that produced relativity large quantitative of Short lived green house gases, as did many developing countries. On the issue of sinks, the CSE argued, the natural sinks ought to be allocated such that each nation was credited with its entire national sinks as well as a share of the global sinks based on its share of world population. And the CSE also criticized, third world representatives who had unquestionably accepted the statistics and analysis of the WRI. Thus the CSE report impressed the Indian government and provided the negotiations with some ammunition to attack the WRI report.TERI's main contribution was technical evaluations of the costs of responding to climate change; its confirmation that these costs were high reinforced the government's determination to avoid them and to obtain assistance from the North.

In general, most NGOs preferred to concentrate their energies on domestic situation and ignored the government's foreign environmental issues because of the scale of domestic environmental problems, such as lack of clean drinking water and soil erosion. And merely in terms of participation in international meetings, prominent NGOs like CSE and TERI often lacked the funds to sponsor the participation of their representatives. Further more lack of access to information, also constrained the ability of NGOs to influence government policy on global environmental issues. The media's influence over government policy was also minimal as a leading environmental journalist admits. Ozone depletion was the only issue to receive a certain amount of critical media coverage, particularly in the early stages when the government's lack of policy was evident. As far as the climate change issues were concerned, there was hardly any sustained, critical commentary in the media; the only source of information for NGOs was government. But the government consistently maintained that it could not be expected to go out of its ways to provide information to NGOs.

Thus, one thing was clear particularly with climate change issue that the government did not wish to encourage public debate lest this divert attention from other pressing problems. The government had no resources to allocate to the climate issue, its entire environmental budget already proving insufficient to cope with existing environmental problems. The most it was prepared to advocate was 'no-regrets' domestic policy. Taking actions to deal with climate change only if they could be justified on other ground too. Thus explanations regarding the limited influence of NGOs on Indian policy must also take account of the government's secretiveness with regard to its policy. The government did not keep NGOs informed on a regular basis about the progress of negotiations, nor did it generally consult them on policy. This behaviour was primarily the result of the bureaucracy's traditional inclination, to restrict access to policy making to as few outsides as possible, successive environment ministers, moreover, reposed total trust in bureaucracy, and did little to relax the latter's grip on policy-making.

Besides, NGOs and the media, business interests constituted another group, which had very limited impact on government policy. This was due to the indications that the impact of policies under the climate convention on business community in India would be negligible. Greenhouse emission curbs were countries, simply not on the agenda for developing countries, and the conventional wisdom was that they would not be for several decades. Therefore, while the business community supported the government's policy of rejecting curbs on Greenhouse gas production, and joined calls for the transfer of environmentally friendly and energy-efficient technology from the North, it had little influence as such on the formulation of Indian policy.

Thus limited domestic political interests in global environmental issues and the limited impact of non-state actors left the government with considerable degree of autonomy in policy formulation.

The different global environmental negotiations revealed impressive consistency in the character of Indian interests and preferences, attributable to the construction of Indian policy around traditional foreign environmental policy concepts such as sovereignty, equity, the poverty is (the greatest polluter), orthodoxy and Third World solidarity. These were key criteria employed by the Indian government to assess the outcome of each global environmental negotiations, namely the extent to which India had evaded international commitments and the degree to which the flow of financial and technological resources to India were going to be enhanced. Indian interests and preferences can be seen as encompassing at least two important categories of goals defensive goals to do with preserving sovereignty, ensuring equity, and reducing vulnerability; and more assertive goals, to do with securing, economic benefits and excursing more power in the international system. India's defensive sovereignty concerns can be categorised into these types. That there should be no firm obligation on it; that there should be no international control on it; and that there should be no market interference in the principles governing the operation of India's environmental agreements. India's rejection of firm obligations was reflected in its insistence on its sovereign right to establish its own environmental priorities and its rejection of outside interference in its policies. Indian policy alleviation through rapid economic development was a more pressing priority in the short term than global environmental protection.

Besides the defensive goals Indian policy also encompassed more assertive goals to do with securing economic benefits and exercising more power in the international system. Economic considerations played a very important role in Indian policy, reflecting the traditional prioritisation of development over the environment. Thus in the climate change issue, India opposed any suggestion that it restrain the growth of its green house emission. A sustained increase in energy production and use was to playa controlled role in fuelling its economic development. Although non-conventional and renewable sources of energy and nuclear power to play important parts in this regard, the predominant role was to be that of fossil fuel-bases sources of energy. Particularly petroleum and coal. In addition with a very low per capita consumption of GHG as compared to the world average, and an already substantial commercial deficit, India could well expect to see its emission increase substantially in the future. Consequently Indian policy makers were not prepared to countenance any controls on India's GHG production.

Indian policy also displayed a small but significant desire to exercise more power in the international system. This was reflected in Indian efforts to reduce the control of donor countries over the financial mechanics established under the different environmental conventions. These were partly motivated by equity concerns. But in addition, they also owned to resentment of the control exercised by the North over the lending policies of international financial institutions, notably the World Bank and the IMF. This resentment increased in Indian policy-making, when India was itself forced to obtain World Bank and IMF endorsement of its structural adjustment programme in 1991. Nevertheless, it pressed demands during the different negotiations that each mechanism operate under the authority of the parties to its parent convention, and that the governance of each be equitable balanced between developed and developing countries. It opposed the GEF because of its lack of these features and intimate links with the World Bank.

During the ozone negotiations, when India's economic situation was relatively stable, policy makers were apprehensive about possible trade restrictions that might be imposed by the North if India stayed out of Montreal Protocol. The economic crisis of 1991 increased the apprehensions. Due to the strong backing by North on different global environmental negotiations, India was not inclined to look upon non-participation as a good option. At the same time an incentive for participation was provided by North. The elements constituted a significant source of bargaining power for India and other Southern countries. Thus whereas in the ozone issue India allowed the North to set the agenda and then fought a rearguard action to secure its interests, in the climate change India did not repeat the same mistake and participated in negotiations from the start.

The coordination and alliance of policy with the coalition of developing countries, was the basic strategy that India adopted in global environmental negotiations to secure its interest because of the tradition of Third World Solidarity in Indian Foreign Policy.

In accordance with the importance of Third World Solidarity in its policy, India played a leadership role in articulating Third World concerns and uniting developing countries around a common agenda. India's used to good effect such as the NAM, the Commonwealth in the first and second ministerial conferences of developing countries, on environment and development at Beijing and Kula Lumpur respectively, meetings of the environment Minister of SAARC, G-15 meetings, and a large number of G-77 meetings during the various negotiations on global environmental issues as well as during meeting of UNCED.

India also organised the first significant political conference of developing countries on global environmental issues, the- April 1990 conference of Select Developing Countries on Global Environmental Issues. Other developing countries generally supported most of India's defensive and assertive goals: However, in the climate change issue, there were important divisions within the South over the question of obligations of the developing countries. The AOSIS countries were anxious to see both North and South adopt strong controls on GHG production. The Latin American states tended to favour some obligations, but on a differentiated basis from the North. Most of the African and Asian countries rejected there obligations. These divisions could not be resolved, but were not allowed to disrupt Southern unity over other issues. G-77 statements, meanwhile, tended to carefully avoid elaborating on the question of developing countries obligations.

The majority of developing countries emphasised the North's main responsibility for climate change and were keen to see the North adopt targets to control its GHG emissions. However, the OPEC states did not want such controls to apply to fossil fuels, especially petroleum, least their exports suffer. They therefore kept stressing the scientific uncertainties involved in understanding climate change, and cautioned against participate action. When the EC approved in May 1992 a draft directive to tax energy in order to combat global warming, the OPEC states came out vehemently against the tax.

The OPEC states feared, the tax would reduce the demand for oil and cause investment in oil sector to decline. The differences could not be resolved and undermined to some extent the South's position in the climate change issue. The developing countries were also divided over an appropriate concept to guide the allocation of GHG emission rights amongst states. Developing countries like India and China favoured the concept of per capita equity. Other developing countries, with smaller populations and rapidly growing energy requirements, were ambivalent about the concepts utility.

Oil producing countries like Saudi Arabia were strongly opposed to the concept and thus no agreement could be 'reached on the subject. Thus, despite significant differences over a few of India's goals, the developing countries were generally united in their approval for the others. Hence, India found it relatively easy to generate Third World solidarity in support of most of its interests and preferences. In the ozone issue, the threat of non-cooperation of major developing countries like India and China was credible.

But in the climate change issue, the threat of non-cooperation was not seen as credible option by many developing countries because of their vulnerability to Northern pressure. India, for instance was in a much weaker economic position than it had been during the ozone negotiations, and Indian policy makers, fearing Northern exploitation of India's economic differentiates, stressed moderation in India's bargaining strategy. Division within the South over important issues like the obligations of developing countries, Northern commitments, and the allocation of green house emission rights, also would have reduced the credibility of a Southern threat of non-cooperation. As far as public pressure in the North was concerned, unlike in ozone issue this was counterbalanced by pressure from a small group of skeptical scientists, as well as many industrial groups, economists and others, all of whom were worried by the costs of greenhouse emission reduction strategies and stressed caution and the need for more evidence about the actual nature of threats posed by climate change before any strong steps were taken. Indeed, relative differences in constraints imposed by these groups on different Northern governments also created divisions within the North. Although the South was able to secure some minor concession from the North in climate issue, it was-unable to prevent poor overall outcome, mainly because of its inability to use good sources of bargaining power.

6

Climate Change Factors

Variations in CO_2, temperature and dust from the Vostok ice core over the last 450,000 years. For current global climate change, see Global warming. Climate change is any long-term significant change in the "average weather" that a given region experiences. Average weather may include average temperature, precipitation and wind patterns. It involves changes in the variability or average state of the atmosphere over durations ranging from decades to millions of years. These changes can be caused by dynamic process on Earth, external forces including variations in sunlight intensity, and more recently by human activities.

In recent usage, especially in the context of environmental policy, the term "climate change" often refers to changes in modern climate (see global warming). For information on temperature measurements over various periods, and the data sources available, see *temperature record.* For attribution of climate change over the past century, see attribution of recent climate change.

Climate change factors

This work needs additional citations for verification. Please help improve this article by adding reliable references. Unsourced material may be challenged and removed. (February 2008)

Climate changes reflect variations within the Earth's atmosphere, processes in other parts of the Earth such as oceans and ice caps, and the effects of human activity. The external factors that can shape climate are often called climate forcings and include such processes as variations in solar radiation, the Earth's orbit, and greenhouse gas concentrations.

Variations within the Earth's climate

Weather is the day-to-day state of the atmosphere, and is a chaotic non-linear dynamical system. On the other hand, *climate* — the average state of weather — is fairly stable and predictable. Climate includes the average temperature, amount of precipitation, days of sunlight, and other variables that might be measured at any given site. However, there are also changes within the Earth's environment that can affect the climate.

Glaciation

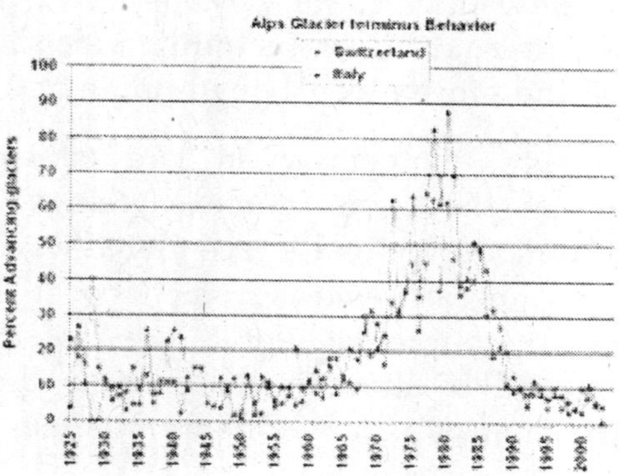

Fig. 5.1

Percentage of advancing glaciers in the Alps in the last 80 years

Glaciers are recognized as being among the most sensitive indicators of climate change, advancing substantially during climate cooling (e.g., the Little Ice Age) and retreating during climate warming on moderate time scales. Glaciers grow and collapse, both contributing to natural variability and greatly amplifying externally forced changes. For the last century, however, glaciers have been unable to regenerate enough ice during the winters to make up for the ice lost during the summer months (see glacier retreat).

The most significant climate processes of the last several million years are the glacial and interglacial cycles of the present ice age. Though shaped by orbital variations, the internal responses involving continental ice sheets and 130 m sea-level change certainly played a key role in deciding what climate response would be observed in most regions. Other changes, including Heinrich events, Dansgaard–Oeschger events and the Younger Dryas show the potential for glacial variations to influence climate even in the absence of specific orbital changes.

Ocean variability

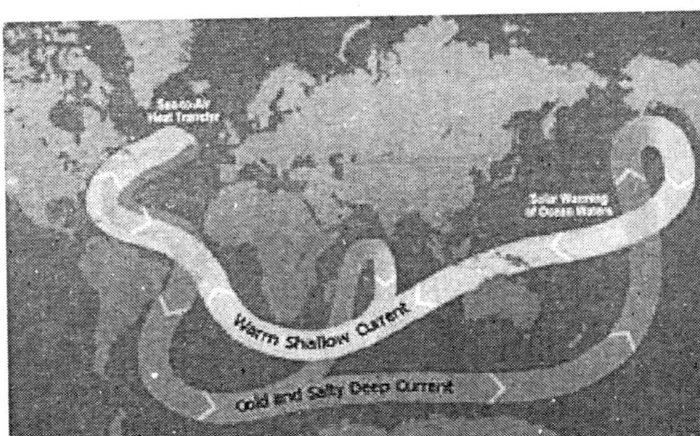

A schematic of modern thermohaline circulation

On the scale of decades, climate changes can also result from interaction of the atmosphere and oceans. Many climate fluctuations — including not only the El Niño Southern oscillation (the best known) but also the Pacific decadal oscillation, the North Atlantic oscillation, and the Arctic oscillation — owe their existence at least in part to different ways that heat can be stored in the oceans and move between different reservoirs. On longer time scales ocean processes such as thermohaline circulation play a key role in redistributing heat, and can dramatically affect climate.

The memory of climate

More generally, most forms of internal variability in the climate system can be recognized as a form of hysteresis, meaning that the current state of climate reflects not only the inputs, but also the history of how it got there. For example, a decade of dry conditions may cause lakes to shrink, plains to dry up and deserts to expand. In turn, these conditions may lead to less rainfall in the following years. In short, climate change can be a self-perpetuating process because different aspects of the environment respond at different rates and in different ways to the fluctuations that inevitably occur.

Non-climate factors driving climate change

Greenhouse gases

Greenhouse gas

Carbon dioxide variations during the last 500 million years

Current studies indicate that radiative forcing by greenhouse gases is the primary cause of global warming. Greenhouse gases are also important in understanding Earth's climate history. According to these studies, the greenhouse effect, which is the warming produced as greenhouse gases trap heat, plays a key role in regulating Earth's temperature.

Over the last 600 million years, carbon dioxide concentrations have varied from perhaps >5000 ppm to less than 200 ppm, due primarily to the effect of geological processes and biological innovations. It has been argued by Veizer et al., 1999, that variations in greenhouse gas concentrations over tens of millions of years have not been well correlated to climate change, with plate tectonics perhaps playing a more dominant role. More recently Royer et al.[1] have used the CO_2-climate correlation to derive a value for the climate sensitivity. There are several examples of rapid changes in the concentrations of greenhouse gases in the Earth's atmosphere that do appear to correlate to strong warming, including the Paleocene–Eocene thermal maximum, the Permian–Triassic extinction event, and the end of the Varangian snowball earth event.

During the modern era, the naturally rising carbon dioxide levels are implicated as the primary cause of global warming since 1950. According to the Intergovernmental Panel on Climate Change (IPCC), 2007, the atmospheric concentration of CO_2 in 2005 was 379 ppm^3 compared to the pre-industrial levels of 280 ppm^3. Thermodynamics and Le Chatelier's principle explain the characteristics of the dynamic equilibrium of a gas in solution such as the vast amount of CO_2 held in solution in the world's oceans moving into and returning from the atmosphere. These principles can be observed as bubbles which rise in a pot of water heated on a stove, or in a glass of cold beer allowed to sit at room temperature; gases dissolved in liquids are released under certain circumstances.

Plate tectonics

On the longest time scales, plate tectonics will reposition

continents, shape oceans, build and tear down mountains and generally serve to define the stage upon which climate exists. More recently, plate motions have been implicated in the intensification of the present ice age when, approximately 3 million years ago, the North and South American plates collided to form the Isthmus of Panama and shut off direct mixing between the Atlantic and Pacific Oceans.

Solar variation

Variations in solar activity during the last several centuries based on observations of sunspots and beryllium isotopes.

The sun is the ultimate source of essentially all heat in the climate system. The energy output of the sun, which is converted to heat at the Earth's surface, is an integral part of shaping the Earth's climate. On the longest time scales, the sun itself is getting brighter with higher energy output; as it continues its main sequence, this slow change or evolution affects the Earth's atmosphere. It is thought that, early in Earth's history, the sun was too cold to support liquid water at the Earth's surface, leading to what is known as the Faint young sun paradox.[citation needed].

On more modern time scales, there are also a variety of forms of solar variation, including the 11-year solar cycle and longer-term modulations. However, the 11-year sunspot cycle does not manifest itself clearly in the climatological data. Solar intensity variations are considered to have been influential in triggering the Little Ice Age, and for some of the warming observed from 1900 to 1950. The cyclical nature of the sun's energy output is not yet fully understood; it differs from the very slow change that is happening within the sun as it ages and evolves.[citation needed].

To quote Spencer R. Weart (The discovery of Global Warming 2003) "The study of [sun spot] cycles was generally popular through the first half of the century. Governments had collected a lot of weather data to play with and inevitably people found correlations between sun spot cycles and select weather patterns. If rainfall in England didn't fit the cycle, maybe storminess in New England would. Respected scientists and enthusiastic amateurs insisted they had found patterns reliable enough to make predictions. Sooner or later though every prediction failed. An example was a highly credible forecast of a dry spell in Africa during the sunspot minimum of the early 1930s. When the period turned out to be wet, a meteorologist later recalled "the subject of sunspots and weather relationships fell into dispute, especially among British meteorologists who witnessed the discomfiture of some of their most respected superiors" Even in the 60s he said, "For a young researcher to entertain any statement of sun-weather relationships was to brand oneself a crank."

Orbital variations

In their effect on climate, orbital variations are in some sense an extension of solar variability, because slight variations in the Earth's orbit lead to changes in the distribution and abundance of sunlight reaching the Earth's surface. Such orbital variations, known as Milankovitch cycles, are a highly predictable consequence of basic physics due to the mutual interactions of the Earth, its moon, and the other planets. These variations are considered the driving factors underlying the glacial and interglacial cycles of the present ice age. Subtler variations are also present, such as the repeated advance and retreat of the Sahara desert in response to orbital precession.

Volcanism

A single eruption of the kind that occurs several

times per century can affect climate, causing cooling for a period of a few years. For example, the eruption of Mount Pinatubo in 1991 affected climate substantially. Huge eruptions, known as large igneous provinces, occur only a few times every hundred million years, but can reshape climate for millions of years and cause mass extinctions. Initially, scientists thought that the dust emitted into the atmosphere from large volcanic eruptions was responsible for the cooling by partially blocking the transmission of solar radiation to the Earth's surface. However, measurements indicate that most of the dust thrown in the atmosphere returns to the Earth's surface within six months.

Volcanoes are also part of the extended carbon cycle. Over very long (geological) time periods, they release carbon dioxide from the earth's interior, counteracting the uptake by sedimentary rocks and other geological carbon dioxide sinks. However, this contribution is insignificant compared to the current anthropogenic emissions. The US Geological Survey estimates that human activities generate more than 130 times the amount of carbon dioxide emitted by volcanoes.[2]

Attribution of recent climate change

Human influences on climate change

Anthropogenic factors are human activities that change the environment and influence climate. In some cases the chain of causality is direct and unambiguous (e.g., by the effects of irrigation on temperature and humidity), while in others it is less clear. Various hypotheses for human-induced climate change have been debated for many years.

The biggest factor of present concern is the increase in CO_2 levels due to emissions from fossil fuel combustion, followed by aerosols (particulate matter in the atmosphere), which exert a cooling effect, and cement manufacture. Other factors, including land use, ozone depletion, animal agriculture[3] and deforestation, also affect climate.

Fossil fuels

Carbon dioxide variations over the last 400,000 years, showing a rise since the industrial revolution.

Beginning with the *industrial revolution* in the 1850s and accelerating ever since, the human consumption of fossil fuels has elevated CO_2 levels from a concentration of ~280 ppm to more than 380 ppm today. These increases are projected to reach more than 560 ppm before the end of the 21st century. It is known that carbon dioxide levels are substantially higher now than at any time in the last 750,000 years.[4] Along with rising methane levels, these changes are anticipated to cause an increase of 1.4–5.6 °C between 1990 and 2100 (see global warming).

Aerosols

Anthropogenic aerosols, particularly sulphate aerosols from fossil fuel combustion, exert a cooling influence[5]. This, together with natural variability, is believed to account for the relative "plateau" in the graph of 20th-century temperatures in the middle of the century.

Cement manufacture

Cement manufacturing is the third largest cause of man-made carbon dioxide emissions. Carbon dioxide is produced when calcium carbonate ($CaCO_3$) is heated to produce the cement ingredient calcium oxide (CaO, also called *quicklime*). While fossil fuel combustion and deforestation each produce significantly more carbon dioxide (CO_2), cement-making is responsible for approximately 2.5% of total worldwide emissions from industrial sources (energy plus manufacturing sectors).[6]

Land use

Prior to widespread fossil fuel use, humanity's largest effect on local climate is likely to have resulted from land use. Irrigation, deforestation, and agriculture fundamentally change the environment. For example, they change the amount of water going into and out of a given location. They also may change the local albedo by influencing the ground cover and altering the amount of sunlight that is absorbed. For example, there is evidence to suggest that the climate of Greece and other Mediterranean countries was permanently changed by widespread deforestation between 700 BC and 1 AD (the wood being used for shipbuilding, construction and fuel), with the result that the modern climate in the region is significantly hotter and drier, and the species of trees that were used for shipbuilding in the ancient world can no longer be found in the area.

A controversial hypothesis by William Ruddiman called the early anthropocene hypothesis[7] suggests that the rise of agriculture and the accompanying deforestation led to the increases in carbon dioxide and methane during the period 5000–8000 years ago. These increases, which reversed previous declines, may have been responsible for delaying the onset of the next glacial period, according to Ruddimann's overdue-glaciation hypothesis.

In modern times, a 2007 Jet Propulsion Laboratory study [8] found that the average temperature of California has risen about 2 degrees over the past 50 years, with a much higher increase in urban areas. The change was attributed mostly to extensive human development of the landscape.

Livestock

According to a 2006 United Nations report, Livestock's

Long Shadow, livestock is responsible for 18% of the world's greenhouse gas emissions as measured in CO_2 equivalents. This however includes land usage change, meaning deforestation in order to create grazing land. In the Amazon Rainforest, 70% of deforestation is to make way for grazing land, so this is the major factor in the 2006 UN FAO report, which was the first agricultural report to include land usage change into the radiative forcing of livestock. In addition to CO_2 emissions, livestock produces 65% of human-induced nitrous oxide (which has 296 times the global warming potential of CO_2) and 37% of human-induced methane (which has 23 times the global warming potential of CO_2).[3]

Interplay of factors

If a certain forcing (for example, solar variation) acts to change the climate, then there may be mechanisms that act to amplify or reduce the effects. These are called positive and negative feedbacks. As far as is known, the climate system is generally stable with respect to these feedbacks: positive feedbacks do not "run away". Part of the reason for this is the existence of a powerful negative feedback between temperature and emitted radiation: radiation increases as the fourth power of absolute temperature.

However, a number of important positive feedbacks do exist. The glacial and interglacial cycles of the present ice age provide an important example. It is believed that orbital variations provide the timing for the growth and retreat of ice sheets. However, the ice sheets themselves reflect sunlight back into space and hence promote cooling and their own growth, known as the ice-albedo feedback. Further, falling sea levels and expanding ice decrease plant growth and indirectly lead to declines in carbon dioxide and methane. This leads to further cooling. Conversely, rising temperatures caused, for example, by anthropogenic emissions of greenhouse gases could lead to decreased snow and ice cover, revealing darker ground underneath, and consequently result in more absorption of sunlight. [9]

Water vapor, methane, and carbon dioxide can also act as significant positive feedbacks, their levels rising in response to a warming trend, thereby accelerating that trend. Water vapor acts strictly as a feedback (excepting small amounts in the stratosphere), unlike the other major greenhouse gases, which can also act as forcings.

More complex feedbacks involve the possibility of changing circulation patterns in the ocean or atmosphere. For example, a significant concern in the modern case is that melting glacial ice from Greenland will interfere with sinking waters in the North Atlantic and inhibit thermohaline circulation. This could affect the Gulf Stream and the distribution of heat to Europe and the east coast of the United States.

Other potential feedbacks are not well understood and may either inhibit or promote warming. For example, it is unclear whether rising temperatures promote or inhibit vegetative growth, which could in turn draw down either more or less carbon dioxide. Similarly, increasing temperatures may lead to either more or less cloud cover.[10] Since on balance cloud cover has a strong cooling effect, any change to the abundance of clouds also affects climate.[11]

Monitoring the current status of climate

Scientists use "Indicator time series" that represent the many aspects of climate and ecosystem status. The time history provides a historical context. Current status of the climate is also monitored with climate indices.[12][13][14][15]

Evidence for climatic change

Evidence for climatic change is taken from a variety of sources that can be used to reconstruct past climates. Most of the evidence is indirect—climatic changes are inferred from changes in indicators that reflect climate, such as vegetation, dendrochronology, ice cores[16], sea level change, and glacial retreat.

Pollen analysis

Palynology is the science that studies contemporary and fossil palynomorphs, including pollen. Palynology is used to infer the geographical distribution of plant species, which vary under different climate conditions. Different groups of plants have pollen with distinctive shapes and surface textures, and since the outer surface of pollen is composed of a very resilient material, they resist decay. Changes in the type of pollen found in different sedimentation levels in lakes, bogs or river deltas indicate changes in plant communities; which are dependent on climate conditions[17][18].

Beetles

Remains of beetles are common in freshwater and land sediments. Different species of beetles tend to be found under different climatic conditions. Knowledge of the present climatic range of the different species, and of the age of the sediments in which remains are found, allows past climatic conditions to be inferred.[19]

Glacial geology

Advancing glaciers leave behind moraines and other features that often have datable material in them, recording the time when a glacier advanced and deposited a feature. Similarly, by tephrochronological techniques, the lack of glacier cover can be identified by the presence of datable soil or volcanic tephra horizons. Glaciers are considered one of the most sensitive climate indicators by the IPCC, and their recent observed variations provide a global signal of climate change. See Retreat of glaciers since 1850.

Examples of climate change

Climate change has continued throughout the entire history of Earth. The field of paleoclimatology has provided information of climate change in the ancient past, supplementing modern observations of climate.

1. Climate of the deep past
   - Faint young sun paradox
   - Snowball earth
   - Oxygen Catastrophe
2. Climate of the last 500 million years
   - Phanerozoic overview
   - Paleocene–Eocene Thermal Maximum
   - Cretaceous Thermal Maximum
   - Permo–Carboniferous Glaciation
   - Ice ages
3. Climate of recent glaciations
   - Dansgaard–Oeschger event
   - Younger Dryas
   - Ice age temperatures
4. Recent climate
   - Holocene Climatic Optimum
   - Medieval Warm Period
   - Little Ice Age
   - Year Without a Summer

- Temperature record of the past 1000 years
- Global warming
- Hardiness Zone Migration

Climate change and biodiversity

The life cycles of many wild plants and animals are closely linked to the passing of the seasons; climatic changes can lead to interdependent pairs of species (e.g. a wild flower and its pollinating insect) losing synchronization, if, for example, one has a cycle dependent on day length and the other on temperature or precipitation. In principle, at least, this could lead to extinctions or changes in the distribution and abundance of species. One phenomenon is the movement of species northwards in Europe. A recent study by Butterfly Conservation in the UK[20], has shown that relatively common species with a southerly distribution have moved north, whilst scarce upland species have become rarer and lost territory towards the south. This picture has been mirrored across several invertebrate groups. Drier summers could lead to more periods of drought[21], potentially affecting many species of animal and plant. For example, in the UK during the drought year of 2006 significant numbers of trees died or showed dieback on light sandy soils. In Australia, since the early 90s, tens of thousands of flying foxes (Pteropus) have died as a direct result of extreme heat[22]. Wetter, milder winters might affect temperate mammals or insects by preventing them hibernating or entering torpor during periods when food is scarce. One predicted change is the ascendancy of 'weedy' or opportunistic species at the expense of scarcer species with narrower or more specialized ecological requirements. One example could be the expanses of bluebell seen in many woodlands in the UK. These have an early growing and flowering season before competing weeds can develop and the tree canopy closes. Milder winters can allow weeds to overwinter as adult plants or germinate sooner, whilst trees leaf earlier, reducing the length of the window for bluebells to complete their life cycle. Organisations such as Wildlife Trust, World Wide Fund for Nature, Birdlife International and the Audubon Society are actively monitoring and research the effects of climate change on biodiversity and advance policies in areas such as landscape scale conservation to promote adaptation to climate change[23].

Climate Change Negotiations from Stockholm to Rio

The twentyfourth UNGA decided to convene a conference on the human environment in 1972 and requested a report from each member country on the state of their environment, a Committee on the Human Environment under the Chairmanship of Pitambar Pant, a member of Planning Commission was set up to prepare India's report. By May 1971 three reports had been prepared: 'Some aspects of Environmental Degradation and Its Control in India', some aspects of problems of Human Settlement in India and some aspects of Rational Management of Natural Resources. As a result of the reports' more stress was put on the need to establish greater coordination and integration in environmental policies and programmes, in February 1972 a National Committee on Environmental Planning and Coordination (NCEPC) was established in the Department of Science and Technology (DST), The NCEPC was an apex advisory body on all matters relating to environmental protection and improvement. The committee was to plan and coordinate, but the responsibility for execution remained with the various ministers and government agencies. It was expected that the success of the NCEPC would depend on the level of cooperation it received from other ministries and departments. The committee was assisted by the DST, and an Office of Environmental Planning and Coordination (OEPC) was set up under the chairman of the committee. Two of the important activities of the OPEC were collaborating with the project appraisal division of the Planning Commission to develop guidelines for evaluating the relative costs and benefits of development projects that would take environmental factors into account, and formulating proposals and coordinating research programmes or environmental problems. Over the time the composition of the NCEPC changed significantly (committee members were appointed for two-year term).While membership of the NCEPC increased from 14 in 1972 to 24in 1977 and 35 in 1979, the number of non-officials decreased. The Committee also became unwieldy, and decision making more complex. Greater bureaucratisation occurred with the addition of more secretaries. The first committee enjoyed some political clout, but this gradually waned.

Until the fifth general election of 1977 none of the political parties in India considered environmental problems worthy of inclusion in their election platforms. The election was won by Congress-I and it immediately set up a committee, chaired by N.D. Tiwari, to recommended legislative measures and administrative machinery to ensure environmental protection. The Tiwari committee's report, submitted on 15 September 1980, made far reaching recommendations, including the creation of Department of the Environment at the federal level. On the basis of recommendations of the Tiwari Committee, a separate Department of Environment (DOE) was established on 1 November 1980. The functions hitherto to performed by the DST were transferred to the newly created depal1ment This institutional development had its immediate nationwide impact on various state governments and the union territories followed the federal example.

Following another recommendation of the Tiwari committee the NCEPC had been replaced by a National Committee on Environmental Planning (NCEP), with functions similar to those of its pre decessor, the NCEP was formed in April 1981 and authorized to prepare on annual 'state of the environment' report. In January 1985 the government decided to upgrade the DOE to ministry level by reconstituting it as the Ministry of Environment and Forests (MOEFV

Growing international concern about global environmental issues in the late 1980s motivated the MOEF in India to constitute an Expert Advisory Committee (EAC) on global environmental issues in July 1989, to advise the government on all aspects related to global warming under the chairmanship of CSIR, Director General Dr. A.P. Mitra.

In May 1990, at its second meeting, the committee reported that with regard to the major greenhouse gases, 'no database exist in the country, nor is there any present programme for the systematic study/collection of data. The MOEF, in an effort to remedy the situation concluded several studies. These included a study initiated in 1989, and expected to take at least two years to complete, on the impact of sea level rise on Indian coastal areas, and a National Methane campaign launched in 1991 to measure India's methane gas production." The CSIR too, initiated a series of scientific repolls on Indian measurments of global phenomena in 1991, devoting the first to a preliminary assessments of greenhouse emissions in India.

Significantly, the Group of Seven *(G-7),* annual summit of Paris states did not address either the question of responsibility for climate change or the question of the division of the cost of global response, on the common efforts to limit emissions of carbon dioxide.

Thus at the meeting of the NAM in September 1989, the then Indian Prime Minister Rajiv Gandhi, threatened Southern non-participation in global efforts to protect the environment by saying that those with inadequate capacities to pay for environmentally sound technologies would be left with no alternative but to let development proceed without due regard for the environment. Rajiv Gandhi proposed a Planet Protection Fund (PPF) to assist developing countries in acquiring technologies at reasonable cost.

Yet, despite the concerns voiced by India, the developing states as a group, failed to put forward a coherent agenda for discussion 'With the North. There were several reasons:

First, although there was general support for the points raised by India, there was less clarity about the emphasis to be placed on different aspect of climate change issue. Some developing countries, notably the small island states, were very concerned about the effects of climate change, especially sea level rise- which could threaten their vel)' existence. To publicize their fears, these countries organized Small States Conference on Sea Level Rise in Maldives in November 1989, followed soon after, on an initiative of Trinidad and Tobago, by the formation of the Alliance of Small Island States (AOSIS). Countries like India were concerned more with costs of strategies to cope with climate change.

Thus, the then Indian President. R. Venkatraman for instance, argued in a speech in November 1989 that the developed countries contributed to global environmental degradation through excessive resource consumption and large scale industrialization intended to support their life styles, and were primarily responsible for adverse changes in the atmosphere and the oceans.

The Indian policy was elaborated in more detail at the New Delhi conference of Select Developing Countries on Global Environmental Issues in April 1990. The conference was organized by India in an effort to strengthen southern solidarity at a crucial stage in the ozone negotiations, and also to highlight the linkages between the different global environmental issues and need for coherence in the South's strategy with regard to each.

In the absence of hard scientific data, Indian scientists approach to the policy making process with regard to climate change followed the pattern observed with Qzone depletion. Though India did not participate in the 1985 Vienna Conference. As a very small producer offices, with less than 0.5 percent of world production, India did not feel its interests would be affected by the decision taken at Vienna.

In the post-Vienna period policy makers continued to believe that ozone depletion was mainly the concern of the developed countries. India, therefore, did not participate in the preparatory meetings for the Montreal Conference in 1987. It sent an observer, without any negotiating powers to represent it at Montreal. The comprehensive measures taken at Montreal, especially the restrictions on trade with non-parties, caught the Indian government off-guard. For this lapse, it received considerable criticism from domestic press. The Hindu (national daily), for example, criticized the government's 'inaction at Montreal' and called it a 'poor commentary on the Government's appreciation of a grave environmental issue. However, the single most important consequence of the ozone issue for India and other developing countries was the lesson it taught them of the importance of being present, at the creation. The ozone issue demonstrated to the developing countries the importance of participating in international negotiations from their very inception, and thus exercising influence over the emergent agenda. Thus in case of Climate Change India did not make the same mistake and participated in international negotiations right from the statl, and in strength since 1990.

Thus the Indian government's perceptions about climate

change issue were outlined in a paper prepared for the conference.

The government argued that:

(a) It is the developed countries, which have created and continue to add to the threats of climate change and it is primarily their responsibility to reverse the situation by setting limits on their emissions of green house gases.

(b) Developing countries contribute little to the problem though their share is increasing. Their resource is scarce and they do not have ready access to technologies required. They need technical and financial assistance to adopt environmentally benign technologies.

(c) And further it argued that there are many factors contributing to climate change and the range of response is wide. Developing countries would accept particular responses only if they do not impede their economic development or reduce the resources currently available for such development.

India's approach thus reflected traditional concern about sovereignty, equity and the importance of economic development. India stressed the conventional wisdom, voiced at conferences like Toronto in 1988 and Noordwijk in 1989, that the North was mainly responsible for green house gas production. India obtained general sUppOl1 at the conference for a narrowly defined position that effectively shifted all responsibility for tackling climate change to the North.

Further more, 'any convention on climate change must provide for technology transfer to the developing countries and funds to meet their resources needs. This was reinforced by reports commissioned by MOEF which indicated that the costs of greenhouse emission limitation strategies were likely to be very substantial, and that many of the technologies required for such strategies were unavailable in India.

The defensive aspect of Indian policy, of avoiding costly international obligations and technological assistance from the North, was to be a constant feature during the negotiations over climate change. India's arguments also reflected its traditional reluctance to divert scarce resources from development purposes towards meeting the costs of tackling climate change. Thus by fixing the primary responsibility for climate change on the North, many of the costs could be avoided.

Thus Developing Countries were to make constant references in the course of negotiations over climate change to Resolution 44/ 228 of UNGA. The General Assembly Resolution 44/228, on the organization of UNCED,21 addressed the question. of responsibility. The resolution affirmed that 'the responsibility for containing, reducing and eliminating global environmental damage must be borne by the countries causing such damage, must be in accordance with their respective capabilities and responsibilities. In particular, it noted that 'the largest part of the current emission of pollutants into environment .... originates in developed countries, and therefore those countries have main responsibility for combating such pollution.

The resolution also recognized that 'new and additional financial resources will have to be channeled to developing countries in order to ensure their full participation in global efforts for environmental protection'.

The Preparatory meeting of Negotiation on Climate

At the UNEP/WMO preparatory meeting for negotiations

on climate convention in September 1990 the developing countries reiterated their opposition expressed at the UNEP Governing council meeting to negotiations under the auspices of UNEP and WMO. They felt the IPCC process had been dominated by experts from North, and had been manipulated to their advantage. The handling of the IPCC plenary session, moreover, had done little to reassure them that their interests would be protected. Finally, the developing countries were nervous about Director of UNEP, Mostafa Tolba, to conclude a convention before the Rio Summit scheduled for 1992, as they felt their interests might take secondary importance to the speedy conclusion of an international agreement. Also, the developing countries pressed for a forum duly established by UN General Assembly to conduct negotiations.

Thus, the preparatory meeting revealed the determination of the developing countries to ensure their interest were protected at this early stage of climate negotiations.

Mostafa Tolba called on industrialized states to help poorer ones. The UN General Assembly was urged to establish formal negotiations towards a framework convention on climate change.

The conference reiterated the need for developing countries to avoid potentially disastrous course followed by industrialized countries in the past, and to adopt modern technologies early in the process of development particularly in regard to energy. The Geneva Declaration pledged scientific and technological expertise, capacity building and easy access to technology and financial resources to help developing nations. The debate on responsibility was led by a 1990report by Washington based NGO, World Resources Institute (WRI), which showed that the annual Greenhouse Gas (GHG) emission of the developing countries almost equaled those of the industrialized world. It claimed that the South's emissions were growing and would overtake the North in the near future.

Developing countries, nevertheless, sought recognition in the conference's ministerial declaration for the main responsibility of industrialized countries for actions to reduce GHG emissions, as past and present emissions of GHGs originate largely in these countries. They also insisted on new and additional financial and technological resources as a condition of their participation. North South compromise was built around 'Common but differentiated responsibility'

The SWCC saw the beginning of differences between developing countries on the climate issue. The Organization of Petroleum Exporting Countries (OPEC) did not want to talk about CO, emissions from fossil fuel use, which accounts for more than half of global warming. The poorest countries, meanwhile, clamored for action, fearing they would suffer most in a warmer world. The small island states, which were particularly vulnerable to the effects of climate change, repeatedly clashed with OPEC, which cited scientific uncertainties about climate change to oppose strong action to control GHG emissions. The majority of developing countries adopted position between these two groups, generally favoring strong action by the North but rejecting such action for themselves.

The General Assembly on 21 December 1990, established the Intergovernmental Negotiating Committee (lNC) for a Framework Convention on Climate Change (FCCC) in Resolution 45/212, entitled 'Protection of Global Climate for Present and Future Generations, under its own auspices but supported by UNEP and WMO. A special voluntary fund was established to help developing nations, especially small island states, to participate in the negotiations.

The US Government hosted the first session of the negotiations, at the Westfields International Conference

Center in Chantilly near Washington DC, from 4-14 February 1991. It elected a bureau, approved special efforts to encourage participating by developing countries, established rules of procedure, and finalised the mandate of the working groups. Thus during the preparatory meetings, the question of representation on the bureau of the negotiating body was stressed to keep the track and influenced the process, India sought membership and was elected as one of the vice-chairman, representing the Asian group. The session also, approved for establishment of special voluntary fund to the support the participation of developing countries in climate talks. The need arose because of the poor representation by the developing countries. Finally, the Washington session established two working groups of INC, the first dealing with 'commitments' related to greenhouse emissions controls and the provision of financial and technological assistance to developing countries, and the second dealing with 'mechanism' to implement the commitments undertaken by states.Two important developments took place before the second session of the INC in June 1991; which were to have significant impacts on India's policy. These were: the exposure of critical weaknesses in India's economy; and the production of an influential report by the Delhi-based NGO, the Centre for Science and Environment (CSE), which challenged the statistics and analysis of 1990 WRI report.

The critical weaknesses in Indian economy were mainly due to three major factors which had been building up over a number of years. These were: inward-looking trade and investment policies; extensive bureaucratic controls over production, investment and trade; and poor economic performance of public sector. There was an, explosive growth in foreign borrowing and government spending during the 1980s. India was forced to take TMF loans worth $ 1.8 billion in the month of January, 1991; when India sought for multi-lateral loans, it entailed it for firm commitments to control and reduce the budget deficit and to undertake structural reforms. Thus the Indian negotiators feared that India's economic vulnerability and its dependence on the North's aid would be exploited by the North. The second development which took place before INC-2 was the criticism of 1990 WRI repoll by the CSE. The CSE criticized the WRI for focusing on current annual emissions, rather than cumulative emissions. The report restarted that, the WRI exaggerated the South's responsibility for climate change, it also pointed that WRI was focusing on short term heating effect and ignoring the long term effect of GHG. It has criticized Third World representatives who had unquestionably accepted the statistics and analysis of WRI. The CSE provided Indian negotiators with ammunition to attack WRI report.

At INC-2, the Indian government handed 'non paper' a full draft text proposal for a convention promoting another major concern of developing countries: to ensure that the funding mechanism of the convention would not be in the hands of Northern dominated institution. The Indian government proposed a separate Climate Fund as a funding mechanism for the convention, to be financed by contributions from the developed parties to the convention and to provide finances on a grant basis to developing countries to adapt to and mitigate the adverse effect of climate change. The Indian text proposal for a convention was generally welcomed by developing countries, for projecting North's primary responsibility, finances for South, and democratic administration of any funding mechanism. Despite the general approval division within the South also extended to other issues. They wanted new and additional financial resources to cover their incremental costs and technology on concessional terms and insisted on a dedicated 'Green Fund', which would be controlled by UN rules, rather than the more restrictive practices of the World Bank. The AOSIS states were opposed to the preparation of common text by G-77, as they were keen to have a commitment by all countries to tackle climate change. India and China opposed the designation of the Global Environmental Facility (GEF) as a financing mechanism in the FCCC and pursued for new sources of finance (GEF was set up in 1991 and is managed by World Bank, the UNDP and UNEP).

A proposal by the UK and Japan at INC-2 for a 'pledge and review' process was widely seen as an effort to accommodate the US. Under this process all countries would commit themselves to establishing policies and strategies to limit their GHG emissions. They would report on measures taken by them, and these reports would be the subject of formal review. Specific commitments, such as stabilization by the industrialized countries of their carbon dioxide emissions at 1990 levels by 2000, or commitments by all countries to curb deforestation would be the subject of separate protocols.

India opposed the 'pledge and review' proposals on the grounds that it would impose legal obligations on the developing countries. India feared that the review process could evolve into an intrusive mechanism through which the North would be able to interfere in the national policies of developing countries, especially in important sectors such as energy and industry. However there were differences between the North and the South, over the nature of the financial mechanism, which would transfer resources, and the terms on which technology transfer would take place. India and China also opposed the idea of the joint Implementation (JI) introduced by Norway & Germany to meet the 1990 commitments set in Toronto. Under 11 industrialized countries would provide funds and better technology to cut emissions in developing countries in return for 'credits' to increase their own emissions. These were to be discussed in more details at INC-3.

INC-3 witnessed little changes in the industrialized countries' position. The northern countries, still insisted on universal obligations, and attempted to use the 'pledge and review' concept to get southern countries on board - the developing countries were asked to accept limited binding commitments from the start, instead of common but differentiated responsibilities principle, and then to have periodic reviews to tighten commitment targets.

This provoked a strong reaction from the South, especially India. India was however, willing to accept that developing countries implement specific projects to reduce their GHG emissions, provided the incremental costs are borne by North. India also warned the North of any attempt made by them to pressurize or impose conditions on the southern nations to change their policies would be inadmissible interference in their internal affairs.

Though India's strong response was welcomed by many developing countries, differences within G-77 persisted over aspects of India's position. They insisted on preferential, concessional and non-commercial technology transfers and provision of financial resources by Northern nations. Thus on the matter of financial resources; Developing Countries wanted a separate financial mechanism under the authority of the parties to the convention, while industrialised countries wanted GEF to serve as the mechanism. Both agreed that whatever the mechanism, the climate convention should provide the policy direction. India and other developing countries wanted a separate, democratically administered fund, because that would provide an equal voice to beneficiaries and donors, and fund activities according to the priorities of recipient countries. GEF failed on both counts - it was perceived to follow World Bank operating procedures and was dominated by donor countries. The developing countries also, wanted contributions to the climate fund to be made obligatory, while contributions to

GEF were voluntary.

Shortly before the INC-4, an Organization for Economic Cooperation and Development (OECD) minister's meeting in Paris was held from 2-3 December 1991. The OECD ministers committed themselves to the enhancement and protection of GHG sinks and limitation of GHG emissions especially CO_2. The OECD countries also acknowledged the need to provide financial resources to developing countries. And most of the OECD countries now showed apparent willingness to restructure GEF. On the matter of technology transfer, they were only willing to expand technology co-operation, facilitate the transfer of technologies and to enhance the capacities of developing countries to use and develop technologies.

The Indian Cabinet met on 3 December 1991, shortly before the INC-4, to consider about India's policy in climate negotiations for the first time. The MOEF highlighted the main elements in India's negotiating stance, including the concept of per capita equity, and its opposition to any international review of policies of developing countries, its willingness to consider contractual commitments and demand for separate fund under the climate convention.[54] The cabinet gave its full approval to the policy note by MOEF.

At the INC-4 a greater consensus among the Southern nations began to emerge. They reiterated the 'main responsibility' argument, which the North continued to reject. Many G-77 countries were only willing to concede contractual commitments for specific projects an offer India had already made at previous session. India with 43 developing countries including Brazil, China offered to consider taking feasible measures to address climate change, provided that the full incremental costs involved of adequate and additional financial resources are met by new provision from the industrialized country parties. North south differences persisted over the question of technology transfer, though they had narrowed down somewhat with the North's indication in OECD policy statement of its willingness to restructure GEF.

A consolidated text had been prepared at INC-4, but was replete with evidence of the wide differences that still existed between the North and the South. It became inevitable for the conclusion of negotiation to resume for an additional 9 days from 30 April to 8 May 1992. The G-77 and China were extremely dismayed by the slow rate of progress.

At a special session of GEF in Geneva in February 14, 1992 Chairman of GEF, Mohamed EL-Ashry, agreed that GEF could be modified to service the climate convention. Though many nations still supported for a separate fund, the trend appeared to be acceptance of a modified GEF as the financial mechanism for the FCCC. The US continued to oppose any compromise on the demands for technological and financial resources for the south. Corresponding with the minimal commitments it sought for itself, it did not place any major obligations on the south.

Thus prior to concluding session of INC, developing countries met and reaffirmed their positions on the climate negotiations at several meetings before the final INC session. These included the conference of environment ministers of South Asian Association for Regional Cooperation (SAARC) in New Delhi from 8-9 April, 1992, and the second ministerial conference of developing countries on environment and development in Kuala Lumpur from April 26-29, 1992.

When lNC-5 resumed in April] 992, lNC chairperson Jean Ripert presented a text to the session as the basis for negotiation. The text attempted to limit the convention to 'essentials' and made no mention of stabilization of emission levels by 2000 at 990 level as has been included in the text at the first session of INC-5. Ripert, moreover, unofficially declared the commitments to be non-negotiable, because any revision would be unacceptable to the US.

The negotiating text obliged developing countries to cooperate with North, by providing for the establishment of a Subsidiary Body for Implementation (SBI) to review information relevant to the implementation of the convention. It only committed the North to take all practicable steps to promote, facilitate and finance, as appropriate, access to and transfer of environmentally sound technologies and know how.

India initially described the draft text as 'Linguistic Striptease' which 'dilutes and distorts previous formulations of the specific commitments language', and called on developing countries to stay out of the framework convention if significant changes were net made. This proposal did not receive any support. China in particular, showed no inclination to boycott the convention. Finally India had to satisfy itself with some revisions in the chairperson's text, with support from other developing countries on specific issues. An Indian amendment to SBI was accepted, confining the body to only reviewing the commitments of industrialised countries. Information provided by developing countries would not be considered on a country-by-country basis, but would be assessed in terms of the overall aggregated effect of the steps taken by the parties in the light of the latest scientific assessments concerning climate change.

The convention text presented at Rio de Janeiro, June 3-14, 1992, committed the West to no more than what one country, the US was willing to commit. George Bush, the US president, who had initially refused to come to Rio, if the climate convention did not meet with the US approval, agreed to come once all references to make stabilization targets mandatory were dropped. Industrialized countries had only accepted the 'common but differentiated responsibilities' principle, which was a very diluted version of the polluter pays principle.

The Indian delegation seemed easily pleased by the few changes in the convention text and turned from complete rejection to approval. "The final outcome really depends on the protocols", Indian minister Kamal- Nath said optimistically.65 While the delegation conceded that no guarantees has been obtained from the North with regard to technology transfer, assessment of financial resources was favorable.

In the view of the MOEF delegate, 'the most important gain in the negotiation was the elimination of articles dealing with review of national policies'.66 The Chief Indian negotiator Chandershekar Dasgupta 'ensured that the obligations imposed on us are minimal and, furthermore, that in all areas there is 'differentiation' between developed and developing countries.' Dasgupta however, conceded that the quantum of financial resources to be provided by the North was indeterminate and in addition 'the precise way in which "full incremental costs" will be "agreed" upon remains to be worked out and upon this will depend the extent of the financial benefits to LIS. The issue of financial mechanism was acknowledged to be a compromise between the North's insistence on the GEF and south's preference for a new mechanism. According to Dasgupta 'the outcome was entirely satisfactory from our point or view.'

Thus, the climate convention did not meet all the demand that India and the other developing countries had made during negotiations. On the positive side the FCCC allowed them (South) to exclude themselves from the purview of the review functions of the subsidiary bodies of the convention. It met the demand for new and additional financial resources. But it did not oblige their demand for a separate climate fund.

Climate Change Negotiations

Official initiation of the conference began with a letter dated 20 May 1968 from the permanent representative of Sweden to the secretary general of the United Nations "on the question of convening an international conference on the problems of human environment:"4o The possibility of such a conference had been previously introduced by the Swedish delegation to the plenary session of the· UN General Assembly on 13 December 1967. The Swedish proposal was referred to ECOSOC for consideration, and the UN Secretariat prepared a short paper outlining the activities of the United Nations Organisations and programs relevant to the human environment. On 30 July 1968 ECOSCO adopted Resolution 1346 (XLV), which requested that the United Nations proceed with plans for a conference[43].

On 3 December 1968 the UN General Assembly took up agenda item 91, "The problems of the human environment" and after generally supportive discussion adopted Draft Resolution, 2398 (XXlU) without opposition, thereby setting in motion the preparatory efforts leading to the conference to be held in Stockholm in the summer of 1972. The objectives of the conference stated in the resolution were to provide a framework for comprehensive consideration within the United Nations of the problems of the human environment in order to focus the attention of the governments and Public opinion on the importance and urgency of this question and also to identify those aspects of it that can only, or at best be solved through international cooperation and agreement. The General Assembly requested the secretary general, in consultation with the Advisory Committee on the Application of Science and Technology to Development (ACASTD), to submit through the ECOSOC to the General Assembly a report concerning the nature, scope, and progress of work being done in the field of human environment, the principle environmental problems facing developed and developing countries, the time and methods necessary to prepare for the conference, a possible date and place for its convening, and finally the range of financial implications for the United Nations.

On 26 May 1969 the secretary general submitted his report to ECOSOC, then convened for its forty seventh session. The secretary general also reported the invitation of the government of Sweden to host the conference in Stockholm in June 1972. Following the endorsement of ECOSOC the report was brought before the General Assembly, which by Resolution 2581 (XXIV) on 15 December 1969 established the Preparatory Committee (PREPCOM) whose work would be essential to the success of the Stockholm Conference· The twenty-seven nations Preparatory Committee together with alternatives and advisers, was too large actually to organise the conference.

Therefore, a special staff was appointed under the direction of Maurice Strong, an official of the government of Canada, who was also designated as a secretary-general of the conference. A number of intergovernmental working groups were organised to develop proposals and bases for agreement on the major items of an agenda. One group was to prepare a draft of a Declaration on the Human Environment (membership identical to the full PREPCOM), and there were additional groups on marine pollution, soils, conservation and monitoring and surveillance·

The first session of the preparatory committee was held at the United Nation headquarters in New York 10-20 March 1970. The committee agreed that its main task would be to assist the secretary -general in the selection of topics and headings, for the conference as well as in the formulation of ideas, suggestions and proposals with regard to the content of its program. At its second session, meeting in Geneva 8-19 February 1971, the Preparatory Committee considered a proposed agenda for the conference culminating in the adoption and signature of a Declaration on the Human Environment. The proposed agenda consisted of six main subjects considerations of which was divided among three main principles committees and which were subsequently the principal working divisions of the Stockholm Conference.

The PREPCOM held its third session in New York in September 1971 with the draft Declaration on the Human Environment being a major item on the agenda. On 22 December 1970 the secretary-general had invited governments to comment on the possible form and contents of a draft declaration that one of the intergovernmental working groups had been assigned the task of preparing for consideration by the full preparatory committee. There was substantial agreement among committee members that the declaration should be a document of universally recognised fundamental principles, recommended for action by individuals, national governments, and the international community. Pre conference approval of the declaration was accomplished at the Fourth Session of the PREPCOM, 6-10 March 1972, at which final arrangements for the conference were ratified.

On 5 June 1972, after more than two years of extensive preparation, the United Nations conference on the Human Environment opened its first plenary session at the Royal Opera House in Stockholm. Two conflicting viewpoints were present. From the perspective of the first, the primary concern of the conference was the human impact on the biophysical environment with emphasis on control of pollution and conservation of resources. The second viewpoint held social and economic development as the real issue. To bridge these differences, the concept was advanced that environmental protection was an essential element of social and economic development and as one of the participants in building the bridge observed, "One development had become a dimension of the problem, it quickly gained prominence."

In addressing the conference, Prime Minister of Sweden, declared that his government attached "the greatest importance to the stress laid in the declaration upon the need for development." He found it "an inescapable fact that each individual in the industrialised countries draws, on the average, thirty times more heavily on the limited resources of the earth than his fellow man in the developing countries." And concluded that "these simple facts inevitably raise the question pf quality, and within countries. Olof Plame blamed industrialised countries for ecological and economic exploration, particularly United States. Indira Gandhi, Prime Minister of India, found poverty the greatest polluter. She declared that "many of the advanced countries of today have reached their present affluence by their domination over other races and countries, the exploitation of their own masses and natural resources. They got a head start through sheer ruthlessness, undisturbed by the feelings of compassion or by abstract theories of freedom equality or justice. Mrs. Gandhi's opinion was widely shared among third world representatives and expressed most violently by China's spokesman. Helena Z. Benitaz of the Philippines emphasised the injustice theme, asserting that; it past age of domination has left in many counties of the so-called third world .... stunted and malformed economics perpetuating to this day the poverty of blighted, stagnant and benighted rural communities".

A part of this preparatory process the key developing countries rapidly evolved their own distinctive approach to the Stockholm agenda. This was most coherently summed up in the 'Founex Report', which was produced by a group of developing country scientists and experts for the Stockholm conference, and was echoed, with remarkable solidarity, in political statements by developing countries in the run up to, and through, Stockholm. Both the report and subsequent statements made clear a profound underlying worry in the developing countries that western concern about the environmental damage wrought by industrialisation could create pressures to slow industrial growth worldwide including the developing world. They were particularly concerned about the possibility of environmentally motivated restrictions on aid, investment or trade policies. In reaction they placed a heavy emphasis on sovereignty, and on the rights of countries to choose their own path of economic development, free from international interference, for environmental or other reasons. All this developmental twist given

by the south to the environmental issue took the North by surprise. The developing countries held that if they were to achieve economic growth in a non-polluting global environment they would need extra funding from the west as well as the modern technologies which would assist economic growth while keeping pollution down. This extra Western help would be justified by the need to compensate them for the costs of meeting higher environmental standards, and should be in addition to existing aid flows· Thus Stockholm declaration was visibly the child of the North-South tensions.

To the extent that the developing nations were able to agree among themselves on the issues of environment and development, they were able to have a decisive influence on events. During the pre-conference preparation period they were able to obtain a definition of environmental problems, which reflected their interests, and to guarantee consideration by the Stockholm conference on the issues of compensation and additionality. Because of their relative unity of outlook and because of the need to include them in any agreement, the developing states were able to win at Stockholm a broadened interpretation of compensation and to have legitimised the principle of additionality. At the 27th General Assembly, they were able to marshal their numerical strength to have their views officially adopted.

This consolidating of the political interests never fully materialised at Stockholm, but became effective in the 27th Session of General Assembly in the establishment of UNEP. The General Assembly accepted the Stockholm recommendations, with the developing countries predominating in the establishment of UNEP and the Governing Council. And the General Assembly instructed the council to give special consideration in the formulation of progress and priorities, which might assist in accelerating the economic development of developing countries without environmental disruption.

The acceptance by the General Assembly of the Declaration of Principles and Recommendations for Action of the Stockholm Conference, however did not guarantee nor make automatic the

implementation of the Stockholm recommendations. The recommendations did not carry the full force of international agreements and did not impose binding obligations on the participating nations. Implementation of the Stockholm recommendations for action depended upon the perceived interests and the cooperation capabilities of sovereign nation states.

The principal accomplishments of the Stockholm conference were two fold: the official recognition of the environment as a subject of general international concern and institutionalisation of that concept in the United Nations Environment Programme (UNEP). Implementing the recommendation of Stockholm for a United Nations

Environment Programme, the UNGA on 15 December 1972, through Resolution 2997 (XXVII), established the necessary institutional and financial arrangements. Thus founding of UNEP represented major shifts in the priority given to climatic issues by international organisations. Climate impacts were 'central concerns'55 as this change led to two developments. First there was a series of UN-sponsored conferences during the 1970s on clilhate related problems. Including the UN world food conference in 1974, the UN Water Conference in 1976, and the UN Desertification Conference in 1977.

Thus substantial cooperative research on potential climate changes began with a conference in Stockholm in July 1974, on the 'physical',56 basis of climate and climate modeling.

The conference was organised by GARP. The recommendation of the WMO executive committee for an increased monitoring of CO_2 was followed up at its session in June 1977, where it set up a research and monitoring project to expand existing monitoring of CO/7 The initiation of world climate programme had been endorsed by the UN Economic and Social Council (ECOSOC) and the UN Desertification Conference in 1977

The developments came to a head when WMO in conjunction

with other UN bodies and lCSU, convened the first World Climate Conference (WCC) in Geneva in February 1979. much of the discussion centered around the debate between the ice-age prophets and those who saw, global warming as being more important in the medium term. It seemed that a consensus was slowly building in favour of the latter view.

The final declaration agreed that:

We can say with some confidence that the burning of fossil fuels, deforestation and change of land use have increased the amount of carbon dioxide in the atmosphere ... and it appears plausible that (this) can contribute to a gradual warming of the lower atmosphere, especially at high latitudes .... It is possible that same effects on a regional and global scale may ... become significant before the middle of the next century.

Later, in June 1979, the Eighth WMO Congress formally established the World Climate Programme (WCP) with endorsement of ECOSOC and WCE. The WCP was the first internationally coordinated programme of research into the world's climate system. The WCP provided the organisational framework within which much climate change research has operated. Pos8ibly more importantly, it organised the Villach conference of 1985, in Austria, which began the process through which global warming became politicised· In October 1985, scientists from 29 countries met in Villach, under the WCP auspices, in order to review the then current state of scientific understanding about the green house effect. Two further workshop were held in Villach and Bellagio, Italy in 1987. 'The Villach Bellagio workshops were significant for two main reasons. Not only were global warming trends confirmed, but also for the first time, policies to respond to climatic change were discussed in detail and agreed·

Alongside these developments, the World Commission on Environment and Development (WCED) presented its report 'Our Common Future' (also known as the 'Brundtland Report' after the Norwegian Prime Minister Gro Harlem Brundtland, who chaired the commission) on 27 April 1987. This was a general report on

environmental degradation and how it related to development issues. The commission had been set up by the UN in 1983. Regarding climate, its report reproduced the recommendations of the 1985 Villach conference, and in particular emphasised the 'urgent' necessity of increasing energy efficiency and shifting the fuel mix towards renewable.

The confidence expressed at Villach was based on a significant growth in both the scope and the complexity of climate research during the 1980's. The most important of these developments included much more realistic models of the atmosphere, and the consolidation of the realisation that other anthropogenic gases (CFCs, Methane, Nitrous oxide, tropospheric ozone) are radiatively important. This had been realised in the mid-1970s by some scientists but it was only by the 1980s that it was widely incorporated into models, or that its policy was realised.

The question of global warming subsequently entered the international agenda in 1988. In June 1988, the Toronto Conference on 'The Changing Atmosphere' was held. With over 300 participants from 48 countries, it was the first major international gathering to focus on global warming. The conference was hosted by Canadian government as response to WCED report.

The Toronto conference was the first major international conference on climate change to bring together representative of states, international organisations and non-governmental organisation. It proposed a world atmosphere fund (WAF) to protect the atmosphere, financed in part by a levy on fossil fuel consumption of industrialised countries. It also urged goverl1lmmts to enact national legislation for the protection of the atmosphere, and encouraged them to work towards global Framework Convention on Climate Change (FCCC).

The Toronto Conference also led to a series of international and intergovernmental conferences about global warming. These continued through to late 1990, and provide a great deal of the pressure and momentum which led to the formal negotiations which

started in 1991. In September 1988, the issue first reached the UNGA, with Malta proposing that climate became part of the 'common heritage of mankind (sic), By December of that year, the General Assembly had passed a resolution, endorsing the establishment of the Intergovernmental Panel on Climate Change (IPCC) and urging that the issue became a priority one, but withdrawing from the 'common heritage' concept towards an assertion that climate change was 'merely a 'common concern' of humanity.

The IPCC was established by UNEP and the WMO in 1988, to assess the scientific information and formulate realistic response strategic for the management of the climate change issue.

The I PCC first met on November 9-11 1988, in WMO office in Geneva. At the meeting it agreed its work programme and what its main tasks were. These were dedicated to be:

I. Assessment of available scientific information on climate change;

II. Assessment of environmental and socio-economic impacts of climate change.

III. Formulation of response strategic.

This panel decided to organise this work through three Working Groups (WG) i.e, WG I (science) organised itselfby commissioning lead authors to cover different aspect of scientific problems global warming and having other member review the work, while WG II (Impacts) and WG HI (Responses) worked through sub groups which looked at particular aspects of the problem and reported collectively.

Prior to this, in November 1988 a World Congress on

Climate and Development was held in Hamburg. This called carbon dioxide emissions to be reduced by 30 percent by the year 2000 and 50 percent by 20 15. It argued for unilateral action from the industrialised nation to start that process of change; a global ban on the production and use of CFCs covered by Montreal Protocol by 1995, and urgent

strategies for reversing deforestation and beginning aforestation programms. This was despite some, notably the Soviet climatologist Mikhail Budyko, claiming that global farming would be beneficial to agriculture and emissions should possibly even be deliberately increased.

The theme of universal cooperation was reiterated at the Summit Conference on the Protection of the Global Atmosphere, held at the Hague in March 1989. It recommended the establishment of new international authority with responsibility to tackle global warming and the final declaration Oilly spoke of a role for the International Court of Justice in monitoring compliance. Climate change were not discussed in comprehensive way in the Hague Conference. More over it suffered from the non-endorsement of the US and Soviet Union.

In July 1989 the Group of Seven (G-7) major industrial democracies 'annual summit was held in Paris, and was widely dubbed the Green Summit1[6]. The summit called for 'common efforts to limit emission *ofCo$_l$'*, and stated that a 'framework or umbrella convention was urgently required.

Later, in November, 1989 a large Ministerial conference on atmospheric pollution and 'climatic change' was held at Noordwijk in Netherlands, attended by representatives from Seventy-two states. This conference's declaration Committed it Signatories to stabilising CO_2 emissions at level to be set by the lPCC in its preliminary report to the

Second World Climate Conference in 1990, 'at the latest by the year 2000. However US supported by Japan and the Soviet Union, were resistant to any further declaration.

"However, divisions between developed and developing countries started to surface already at the second plenary session of the IPee. Developing countries, led by India, Brazil and Mexico, who were uneasy about their relatively poor representation on the Panel and its lack of normal UN procedural practice, took their concern to the UNGA in December 1989. At the August 1990 IPCC meeting, Brazil was very vocal and almost prevented adoption

NOTES AND REFERENCES

1. Royer DL, Berner RA, Park J (2007). "Climate sensitivity constrained by CO_2 concentrations over the past 420 million years". Nature 446 (7135): 530–2. doi:10.1038/nature.

2. Volcanic Gases and Their Effects. U.S. Department of the Interior (2006-01-10). Retrieved on 2008-01-21.

3. Steinfeld, H.; P. Gerber, T. Wassenaar, V. Castel, M. Rosales, C. de Haan (2006). Livestock's long shadow.

4. Amos, Jonathan. "Deep ice tells long climate story", BBC, 2006-09-04. Retrieved on 2008-01-21.

5. Charlson, R. J.; S. E. SCHWARTZ, J. M. HALES, R. D. CESS, J. A. COAKLEY JR., J. E. HANSEN, and D. J. HOFMANN (1992-01-24). "Climate Forcing by Anthropogenic Aerosols". Science 255 (5043): 423-430. doi:10.1126/science.255.5043.423. Retrieved on 2008-01-28.

6. Why cement-making produces carbon dioxide

7. Ruddiman, William (2005-12-05). Debate over the Early Anthropogenic Hypothesis. RealClimate. Retrieved on 2008-01-21.

8. California Warming Attributed to Growth by Mandalit del Barco. Day to Day, National Public Radio. 30 Mar 2007.

9. Ahlenius, Hugo (June 2007). Climate feedbacks. United Nations Environment Programme/GRID-Arendal. Retrieved on 2008-01-21.

10. Climate Change 2001: The Scientific Basis

11. For additional discussion of feedbacks relevant to ongoing climate change, see http://www.grida.no/climate/ipcc_tar/wg1/260.htm

12. Arctic Change Indicators

13. Bering Sea Climate and Ecosystem Indicators

14. How scientists study climate change: Some important research concepts used by scientists to study climate variations

15. Baxter, JM & Buckley PJ and Wallace CJ, eds. (2008), *Marine Climate Change Impacts Annual Report Card 2007–2008*, Lowestoft: Marine Climate Change Impacts Partnership, <http://www.mccip.org.uk/arc/2007/default.htm>

16. Petit, J. R.; J. Jouzel, D. Raynaud, N. I. Barkov, J.-M. Barnola, I. Basile, M. Bender, J. Chappellaz, M. Davis, G. Delaygue, M. Delmotte, V. M. Kotlyakov, M. Legrand, V. Y. Lipenkov, C. Lorius, L. PÉpin, C. Ritz, E. Saltzman and M. Stievenard (1999-06-03). "Climate and atmospheric history of the past 420,000 years from the Vostok ice core, Antarctica". Nature 399: 429-436. doi:10.1038/20859. Retrieved on 2008-01-22.

17. Langdon, PG; Barber KE, Lomas-Clarke SH (August 2004). "Reconstructing climate and environmental change in northern England through chironomid and pollen analyses: evidence from Talkin Tarn, Cumbria". *Journal of Paleolimnology* **32** (2): 197-213. doi:10.1023/B:JOPL.0000029433.85764.a5. Retrieved on 2008-01-28.

18. Birks, HH (March 2003). "The importance of plant macrofossils in the reconstruction of Lateglacial vegetation and climate: examples from Scotland, western Norway, and Minnesota, USA". *Quarternary Science Reviews* **22** (5-7): 453-473.

doi:10.1016/S0277-3791(02)00248-2.

19. Coope, G.R.; Lemdahl, G.; Lowe, J.J.; Walkling, A. (1999-05-04). "Temperature gradients in northern Europe during the last glacial—Holocene transition(14—9 14 C kyr BP) interpreted from coleopteran assemblages". *Journal of Quaternary Science 13* (5): 419-433. John Wiley & Sons, Ltd.. Retrieved on 2008-02-18.

20. Fox, R.; Warren, M.S., Asher, J., Brereton, T.M. and Roy (2007). The state of Britain's butterflies 2007. Butterfly Conservation and the Centre for Ecology and Hydrology, Wareham, Dorset. Retrieved on 2008-01-21.

21. McGuirk, Rod; Bernard Lagan, Joseph Kerr (2007-01-30). Australian Drought. Retrieved on 2008-01-21.

22. Welbergen, J. A.; Klose, S. M., Markus, N. & Eby, P. (2008-02-22). "Climate change and the effects of temperature extremes on Australian flying-foxes". Proceedings of the Royal Society B 275 (1633): 419-425. Royal Society Publishing. doi:10.1098/rspb.2007.1385. Retrieved on 2008-01-21.

23. Biodiversity and climate change. *United Nations Environment Programme*. UNEP-WCMC. Retrieved on 2008-01-28.

Emanuel, K. A. (2005) *Increasing destructiveness of tropical cyclones over the past 30 years.*, *Nature*, 436; 686-688 ftp://texmex.mit.edu/pub/emanuel/PAPERS/NATURE03906.

IPCC. (2007) *Climate change 2007: the physical science basis (summary for policy makers)*, IPCC.

Miller, C. and Edwards, P. N. (ed.)(2001) *Changing the Atmosphere: Expert Knowledge and Environmental Governance*, MIT Press

O.P. Dwivedi, *India's Environmental Policies, Programmes and Stewardship,* (New York: St Martin's Press, 1997), p. 52.

O.P. Dwivedi, "India: Pollution Control Policy and Programmes", *International Review of Administrative Sciences,* Vol., 43, no. 2, 1977, pp. 123-33.

National Committee on Environmental Planning and Coordination, Inaguaral Function (proceedings published by the *Department of Science & Technology,* New Delhi, 12 April 1972, p. I.

Ruddiman, W. F. (2003) *The anthropogenic greenhouse era began thousands of years ago*, *Climate Change* 61 (3): 261-293

Ruddiman, W. F. (2005) *Plows, Plagues and Petroleum: How Humans Took Control of Climate*, Princeton University Press

Ruddiman, W. F., Vavrus, S. J. and Kutzbach, J. E. (2005) *A test of the overdue-glaciation hypothesis*, *Quaternary Science Review*, **24**:11

7

Prevention of Global Warming

Realizing a ubiquitous networking society" and "Coexistence with the global environment" are the twin corporate visions of Matsushita for the 21st century. As part of the Matsushita Group of companies, Panasonic Mobile Communications Co., Ltd. is committed to these ideals. We are aiming to realize a networking society where people will be able to access information anywhere and anytime. At the same time, we are seeking a society that is in coexistence with the global environment.We have set ourselves the mission to develop the cutting-edge technologies necessary for Panasonic Mobile Communications to tackle the key environmental problems which hinder our progress towards achieving sustainable development for mankind. In this respect, we aim to contribute to society through the progress of our mobile communication business as a member of Matsushita Group and we strive to minimize the impact to the natural environment from our products and business activities.We are committed to the comprehensive management of chemical substances, especially those which may impact upon environmental or human health. We achieve this through the application of our chemical substance management guidelines for products and factories. Needless to say, we reduce or eliminate any chemicals that may give cause for concern. Further, we also seek to minimize our environmental footprint associated with materials consumption by active participation in recycling programs for end-of-life products. And in our production process, we continuously promote the prevention of global warming through our energy-efficiency improvements, our materials recycling to save precious resources, and our waste reduction programs. At the same time, through our internal awareness-raising projects for staff & their families, every one of our employees appreciates how important it is to live an "eco-life" both in work and at home. Our on-going projects under Panasonic's "Love the Earth Citizens' Campaign" have focused on simple ways to save energy and reduce waste in daily life. These projects continue to be successful in encouraging staff and their families to bring the same conservation commitment demonstrated in the workplace into their homes. Thus, both as one company and as individual citizens, we recognize that we have a responsibility to deliver a sustainable future for the next generation. It is this motivator which encourages Panasonic Mobile Communications to pursue sustainability by delivering "Quality of Life" through the mobile communication business using the most advanced technologies, whilst minimizing the associated environmental impact at the same time.

Risk Management

Legal compliance is an essential requirement for all our business activities.We aim to minimize the pollution risks based on our guideline in all our products and production process on a global basis.Here is an outline of our Risk Management, especially prevention measures for soil, groundwater and air pollution, and emergency response. Surveys for Soil and Groundwater PollutionThe use of VOCs (volatile organic compounds) such as chlorinated organic solvents was abolished in 1996 globally. Since then, we have continuously carried out surveys for soil and groundwater, and pollution that exceeds the guideline values in the country has not been reported before. The results of the surveys are reported to the local government every year.Matsushita Group issued the "Manual for Preventing Contamination of Soil and Groundwater" in 1991, and the surveys for soil and groundwater is carried out based on this manual. To conduct the reliable surveys, we installed totally 12 monitoring wells for groundwater at three manufacturing sites in Japan and have implemented these observations for heavy metal and volatile organic compound pollutions periodically.Monitoring well for groundwater Sampling at Shizuoka Site

Protection of Air Pollution (Protection of Ozone Layer)Since the ozone layer absorbs UVB ultraviolet light from the Sun, Ozone Depletion (ozone layer depletion) is expected to increase surface UVB levels which could lead to damage, including increases in skin cancer and cataract.We abolished the use of specific chlorofluorocarbon (CFC) compounds in 1994, and abolished specific chlorinated organic solvent such as Ethane ,1.1.1-trichloro- in 1996 by adopting non-cleaning PCB and alternative cleaning agent. However we still have specific coolant chlorofluorocarbon within part of some old refrigerative equipment and the location of which is tracked and monitored. Thereafter, it will be disposed according to the "Law concerning the recovery and destruction of fluorocarbons".

Emergency preparation and responseWe have a manual for emergency preparation and response under the general requirements of environmental management system. Based on that, we execute a periodical drill once a year at each site.When we discard industrial waste include some harmful waste which needed special treatment, we prepare the protection measures to prevent flow out based on a manual every time.

Toward Global Warming Prevention

The Fourth Assessment Report has been released by the Intergovernmental Panel on Climate Change (IPCC), reporting that greenhouse gases due to human activities impact on global warming. The initial commitment period (2008-2012) of the Kyoto Protocol - an agreement which establishes goals for reducing greenhouse gas emissions - is going to become effective next year and Takeda considers that ensuring a steady implementation of these efforts toward the goal of reducing total greenhouse gas emissions is our responsibility for the next generation.Manufacturing pharmaceutical products requires less energy consumption compared to other industries and its percentage of total greenhouse gas emissions is relatively low. However, it is true that during each stage of our operations, including the procurement of raw materials, manufacture, distribution and disposal, we release greenhouse gases into the air, which has an impact on the global warming regardless of the volume. In accordance with global warming, an increase of tropical diseases and emerging infectious diseases has been predicted; therefore, as a pharmaceutical manufacturer, playing a role in the medical services domain, Takeda recognizes the significance of efforts to boost global warming prevention.

Takeda's CO_2 Reduction Scheme

Takeda has developed the 9th five-year energy conservation program for the period 2006 - 2010 and our activities toward efficient energy use have been progressing. The 9th energy conservation program sets a target for reducing CO_2 emissions by 40 percent in fiscal 2010 compared to the fiscal 2005 level and we aim to promote our efforts toward reducing CO_2 emissions by implementing thorough energy conservation measures and converting fuels into those generating fewer CO_2 emissions, as well as improving the level of CO_2 emissions on a specific consumption basis. In addition, we also plan to set a goal for CO_2 emissions on a global scale for the future. Takeda pledges to strengthen our efforts to confront global warming issues on a group-wide basis, in order to realize the goal.

Achievements In Fiscal 2006

Takeda group's CO_2 emissions in fiscal 2006 amounted to 460K tons, a 5.5 percent decrease over the previous year. This is considered to be due to our continued efforts to accomplish thorough energy saving measures; implemented mainly within Japan. To cite some examples, at the Hikari Plant where approximately half the CO_2 emissions of the overall Takeda group have been generated, meticulous energy saving activities have been implemented: a reduction in use of heavy oil by improving the operation efficiency of the operating power generation facilities, a weekly review of temperature and humidity control during nonoperational period and focusing the compass of storage shelves in order to keep raw materials in a certain place, allowing a reduction in the power used for cranes. Takeda has accomplished an overall reduction in CO_2 emissions of approximately 27K tons on a group-wide basis.

Future Efforts

Takeda faces an issue of an increase in CO_2 emissions, along with an increased production volume at global production bases, including Takeda Pharma Ireland Limited, whereby its full-scale operation will shortly commence as the Takeda group's first global bulk compound production base.As part of group-wide efforts to respond to such challenges, Takeda will strengthen the reduction in CO_2 emissions by establishing energy saving measures for the global production bases. In fiscal 2007, Takeda plans the conversion of coal into natural gas at Tianjin Takeda Pharmaceuticals Co., Ltd., which is our production base in China.The Japan Pharmaceutical Manufacturers Association has set its voluntary action plan: "Reduction of CO_2 emissions of pharmaceutical companies in the fiscal 2010 to fiscal 1990 levels." Takeda's performance has been constantly maintained below 400K tons - the CO_2 emission levels for fiscal 1990 and we expect to attain the goal set for fiscal 2010.

Trend of Takeda Group's CO_2 emissionsTrend of Takeda's CO_2 emissionsCalculation Method

- **Calculation object**CO_2 emissions for calculation object refer to direct emissions generated by combustion of fossil fuels and indirect emissions from electricity use.
- **CO_2 emissions factor**As for the results in Japan, they are calculated based on the "Law Concerning the Rational Use of Energy" and the CO_2 emissions factor for purchased electricity is used from default value (0.000555 t-CO_2/kWh) designated by the ministry ordinance concerning calculation of greenhouse gas emissions associated with business activities by specific generator. As to the CO_2 emissions factor for purchased electricity outside Japan, country-specific factors are used from the GHG Protocol.[*] Previous data has been amended according to the change in the calculation method.

REFERENCES

Environment Bulletin,Newsletter of World Bank Environment Cell, Washington, D.C., Vol. 4, Fall 1992.

Global Environmental Facility, Report by the Chairman to the May 1993 Participants Meeting, (Washington D.C.The World Bank,1994.)

Global Environmental Facility: Independent Evaluation of the Pilot Phase,Washington D.C. UNDP,and the World Bank, 1994.

Human Development Report,UNDP, (New York: Oxford University Press, 1984).

Monthly Report on World Bank-Implemented Environment Facility Operations, (Washington, C/D.C. The World Bank,November-December 1993.)

8

Industrial Use of Materials

Oxygen Cycle

Oxygen, like carbon and hydrogen, is a basic element of life. In addition, in the form of $O_{3,}$ ozone, it provides protection of life by filtering out the sun's UV rays as they enter the stratosphere. In addition to constituting about 20% of the atmosphere, oxygen is ubiquitous. It also occurs in combination as oxides in the Earth's crust and mantle, and as water in the oceans.

Early in the evolution of the Earth, oxygen is believed to have been released from water vapor by UV radiation and accumulated in the atmosphere as the hydrogen escaped into the earth's gravity. Later, photosynthesis became a source of oxygen. Oxygen is also released as organic carbon in CHO, and gets buried in sediments. The role of oxygen in life is describe in the unit on *Biological Systems*.

Oxygen is highly reactive. A colorless, odorless gas at ordinary temperatures, it turns to a bluish liquid at -183° C. Burning or combustion is essentially oxidation, or combination with atmospheric oxygen. A very broad overview of oxygen cycling in nature. The environment of oxygen in numerous reactions make it hard to present a complete picture.

Oxygen is vital to us in many ways (beside the most obvious—for breathing). Water can dissolve oxygen and it is this dissolved oxygen that supports aquatic life. Oxygen is also needed for the decomposition of organic waste. Wastes from living organisms are "biodegradable" because there are aerobic bacteria that convert organic waste materials into stable inorganic materials. If enough oxygen is not available for these bacteria, for example, because of enormous quantities of wastes in a body of water, they die and anaerobic bacteria that do not need oxygen take over. These bacteria change waste material into H_2S and other poisonous and foul-smelling substances. For this reason, the content of biodegradable substances in waste waters is expressed by a special index called "biological oxygen demand" (BOD), representing the amount of oxygen needed by aerobic bacteria to decompose the waste. The result of not meeting the oxygen demand is described in the section on the *water cycle* (needs anchor to exact spot).

Industrial Use of Materials

"If man began with speech, and civilization with agriculture, industry began with fire. Man did not invent it, probably nature produced the marvel for him...He put the wonder to a thousand uses. First, perhaps... to conquer his fearsome enemy, the dark; then...for warmth,...then he applied it to metals, tempering them, and combining them into stronger and suppler than those in which they had come to his hand..."

Beginning with fire, "industry" or, in these early days, humans learned to use energy to manipulate materials, transforming them to suit our purposes. These rearrangements, while bringing great progress, has also left us with large prices to pay because of our disregard or ignorance of evolving the material use on a scale of time and space that learned its lessons from natural material cycles.

The natural material cycles described in the previous sections are a part of our ecosystem. Over billions of years, materials, energy and life have all evolved with mutual interactions to become part of a natural ecology. Most materials that form part of the biosphere occur in cycles, especially those that play central roles in biological systems.

The use of materials by humans has changed over time in quantity and quality, and especially over the last century. The early use of materials for food, shelter and energy required small amounts of materials for each person and were conserved to large extent because it was hard to get, shape and work with materials. Often, materials remained as part of a product such as a tool or a plough for decades and were reused or recycled.

The trends of material use in products have changed significantly with the technological age - particularly in the 20th century. Plastics are perhaps the most radical material invention, in the way that material permeated society. In 1900, 75% of products used renewable materials (that is, agricultural and forest materials, such as wood and natural fibers). By 1980, 70% of materials used were from non-renewable sources such as ores, minerals, and petroleum. In 1955, 8% of materials in products were petroleum-based (plastics). By 1980, 32% of materials in products were petroleum-based.

Wood and stone were the earliest "materials" to be shaped and used by humans. In the course of time, natural materials such as clay and mud were used for cooking and building. With the advance of technology, discoveries of ways to extract metals from ores, methods to shape and transform materials, and eventually with the knowledge afforded by synthetic, organic, and nuclear chemistry, we have made numerous combinations of materials that have never existed in nature. The age of "synthetic" materials started with forming alloys of metals. Bronze (an alloy of tin and copper), brass (an alloy of zinc and copper), and gold alloys (14 carat gold is 14 parts of gold and 10 parts of copper; 24 carat gold is pure gold metal) have all been used since ancient civilization.

The first transformations of materials gave humans the ability to shape materials into forms they wanted. As pure gold is too soft and pliable, the addition of copper was an early invention to make gold stronger. Metallic elements were alloyed to increase strength to make strong and lasting tools, structures and weapons. Iron, copper, sulfur and phosphorus were probably the materials that received the greatest attention for these uses as "industry" emerged. Of course, stones of various kinds, including marble ($CaCO_3$), granite, gypsum ($MgSO_4$), sand (SiO_2) were all used since early times for building. There were also fuel materials such as wood, coal, and oil, - hydrocarbons - that were used first to build fires, then for controlled extraction of energy.

With increasing technological capabilities, the human ingenuity and art (techne in Greek means art or "cunning" as in Charles Dickens' Artful Dodger) combined with scientific understanding led to unprecedented arrangements and transformations of materials. In pressing to get the functionality of various desirable properties and technological progress in designing and producing new materials, we did not know or think of these materials as part of our natural environmental system. In extracting them from the natural system, and combining them into new forms, we began concentrating materials, producing materials that did not exist in nature, and extracting or "purifying" into elemental form large quantities of elements that existed in nature only in chemical combinations with others.

In these processes, there was no awareness of how much "useless" material - wastes - was produced as we extracted the "useful" material, or of the role that "indestructible" synthetic or even natural materials may play when released into the environment after their use in unprecedented quantities over relatively short periods of time. The greatest ignorance we exhibit as an industrialized "civilization" is perhaps a lack of respect or even a total ignorance - of the role of time and of cycles in providing system balance. In pressing on with our economy, we lost sight of our ecology!

Robert Ayres, one of the originators of the idea of industrial ecology summarizes the crux of our material use in the beginning paragraph of his book

Industrial Ecology, co-authored with Leslie Ayres.

"...every substance extracted from the earth's crust, or harvested from a forest, a fishery or from agriculture, is a potential waste, it soon becomes an actual waste in almost all cases, with a delay of a few weeks to a few years at most. The only exception worth mentioning are long-lived construction materials. In other words, materials consumed by the industrial economic system do not physically disappear. They are merely transformed to less useful forms. In some cases (as with fuels) they are considerably transformed by combination with atmospheric oxygen. In other cases (such as solvents and packaging materials) they are discarded in more or less the same form as they are used. It follows from this simple relationship between inputs and outputs - a consequence of the law of conservation of mass - that economic growth tends to be accompanied by equivalent growth in waste generation and pollution."

Though this realization has not fully hit most parts of our society even today, the 1960's environmental consciousness articulated three features/problems of the industrial rearrangement of material: resource depletion, increasing waste materials, and the presence of toxic materials in the environment. The first idea - resource depletion which came from industry - was the realization that the earth does not have an infinite bank of materials. The second of increasing waste came from people noticing the ugliness of waste - as garbage on highways, junked cars, the "blooms" of algae in lakes from over-nitrificiation from detergents and pesticides, or mounds of mining overburden. The presence of toxic materials in the environment was a slower realization, spurred most notably by Rachel Carsons' Silent Spring which has a chapter titled "And the Birds Sing No More" to signify the impact of the pesticide DDT which weakened the eggshells of birds.

Prior to this century, industrial countries did not consider pollution or resource shortage a serious problem. When resources in the home countries were hard to access, trade or conquests were used to procure material. Historians Will and Ariel Durant make the point that the Industrial Revolution came to England first because of the long history of British command of the seas, because science in England was mainly "directed by men of practical bent", and because England had a constitutional government sensitive to business interests.

The reaction to resource depletion in this century was a foray into ideas of conservation and resource management, finding out how the economies of recycling would compare with the economies of resource extraction from nature. Thus early in this century, the aluminum industry began its quest for recycling. Aluminum recycling is one of the most advanced processes of recycling because of decades of learning. Conservation ideas also developed through the desire to conserve the natural beauty of forests and other landscapes.

In the 1970's, economic and political pressures such as the oil cartel and depletion of coal led to ideas of energy conservation in the United States. Energy-intensive, material-producing industries such as aluminum, and paper started serious looks into recycling. Energy conservation measures in the U.S. also led to down-sizing of cars, which had hitherto gone on with the philosophy of "bigger the better" - station wagons for family uses and recreational vehicles for vacations were a common sight during the economic upsurge of the 1950's and early 1960's. Automobile recycling became big and achieved 90% material recovery in the 1980's.

In 1994, the National Environmental Technology Act was passed to encourage the government to work with industry to promote technologies that will have positive environmental impact. Particularly, the government was interested in and more efficient and non-polluting technologies so that we can maintain our current standard of living in light of population growth. In the Senate, those supporting the measure contended that "coming up with new, nonpolluting technologies will ultimately help efficiency and the economy."

All of these conservation measures were, of course, totally anthropocentric, with an objective of preserving our conveniences, and meeting our wants and desires. It was not the ethos of conservation practiced by older civilizations with a respect for value and an eye to the future - to the "seventh generation" as described in the unit on Ethical Systems.

Faced with resource depletion, we began defining materials as renewable and nonrenewable resources. Materials like coal that can only be replaced over long periods of time compared to the time periods of human activities involving their use, are called nonrenewable resources. Materials such as wood which can be replenished in reasonable periods of time are called renewable resources. Note that these are renewable only if we replant trees at a rate that keeps up with use. Thus the *Redwoods* or old-growths forest trees should not be looked on as "renewable" resources because of the time it would take to replace.

The increasing waste and its disposal was the second mounting problem. Europe with its older industrial base and limited land area felt the waste problem sooner and more acutely than the United States. The concentration

DEFINING THE ENVIRONMENTALLY RESPONSIBLE FACILITY

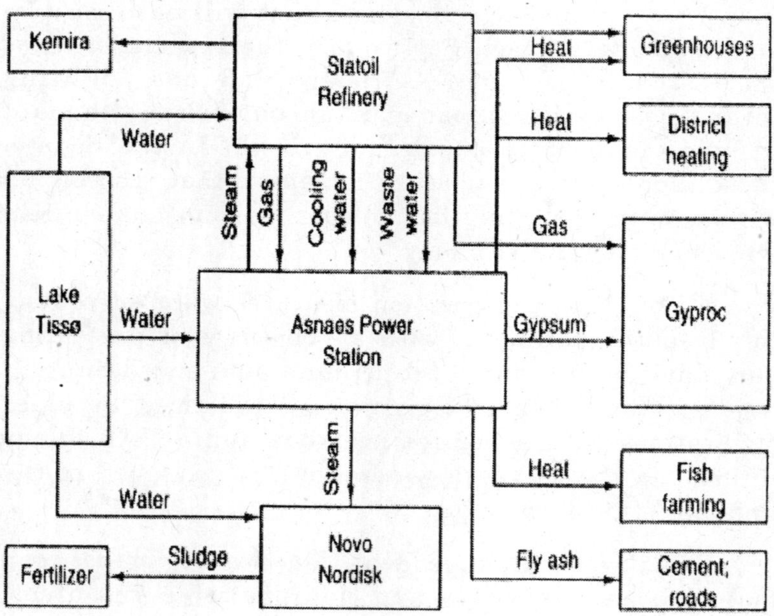

Industrial ecosystem at Kalunborg, Denmark

of people - and hence of quantities of waste is higher in the urban areas. In his book, (link to anobib) The Search for the Ultimate Sink, historian Joel Tarr writes that urban pollution is the "product of the interaction among technology, scientific knowledge, human culture and values and the environment", reduced or exacerbated at times with environmental policy and control technology. Sinking of waste and recycling of usable items have been a part of all cultures. But an economy that valued overproduction and marketed convenience items as necessary to the quality of life and the accompanying perception we have of an inexhaustible resource supply has led to a throwaway society, especially in the U.S.

Often, in analyzing environmental problems, we focus on industrial generation of waste, without a full realization that industry generates waste to meet consumer demands. In 1976, an average consumer spent 15% of his/her income on "durable goods" - automobiles and parts, furniture and other appliances, 18.5% on food, and 22% on "nondurable goods" such as clothing and shoes, gasoline, and alcoholic beverages! (U.S. Bureau of Economic Analysis-link?)

This situation found the U.S. in ridiculous scenarios such as: a boat of garbage from New York City floating in the Hudson River and mountains of Uranium mill tailings blowing radioactive material around some of the western United States. In addition to garbage and sewage, industrial wastes in air, water, and land have also posed formidable problems. The philosophy of waste management has been evolving slowly, spurred by regulation, but making slow progress in the U.S. because of various types of social perceptions.

Toxic materials in the environment is a much subtler problem. Basically, the problem arises when we produce new compounds or isolate elements in configurations not found in nature with properties such as a resistance to degradation. Many of these materials do not decompose on exposure to the usual agents that nature uses to decompose — air, light, water, or bacteria. Persistent substances, such as plastics and chlorofluorocarbons, produce cumulative, long-term and long-range effects on plant, animal and human health.

In many cases the effects were not predictable because of our lack of knowledge and the absence of consideration of systemic, long-term impacts. Joe Thornton writes in (anobib link) Pandora's Poison, a book on the impacts of chlorine compounds, "the chemistry of the chlorine atom gives chlorine gas and organochlorines useful properties, but these same qualities (high reactivity and chemical stability) create enormous environmental problems... Organochlorines that are stable in their intended use, however, are also persistent in the environment." [Thornton].

PRODUCT LIFE CYCLE ANALYSIS

(LCA)

Flow of product through society

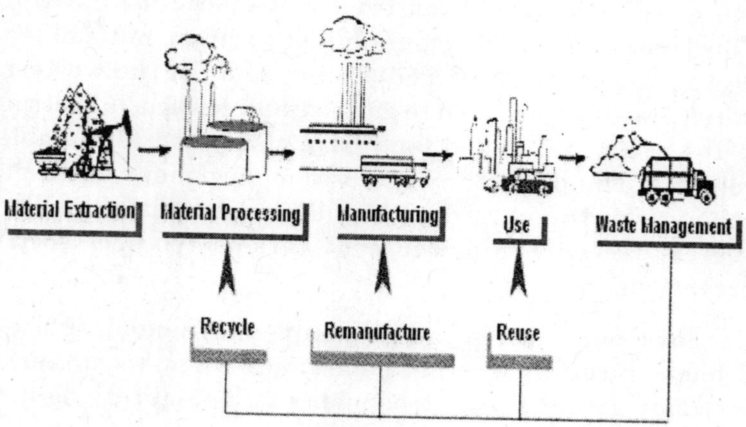

Organochlorines, other organic compounds and metals give rise to health hazards such as cancer and endocrine disruptions discussed in the unit on Heath and Risk.

The U.S. Bureau of Mines maintains an account of the world production and flows of metals and other primary materials such as coal and other fossil fuels. Table shows the amount of some metal ores extracted in 1988.

Aluminum production and recycling

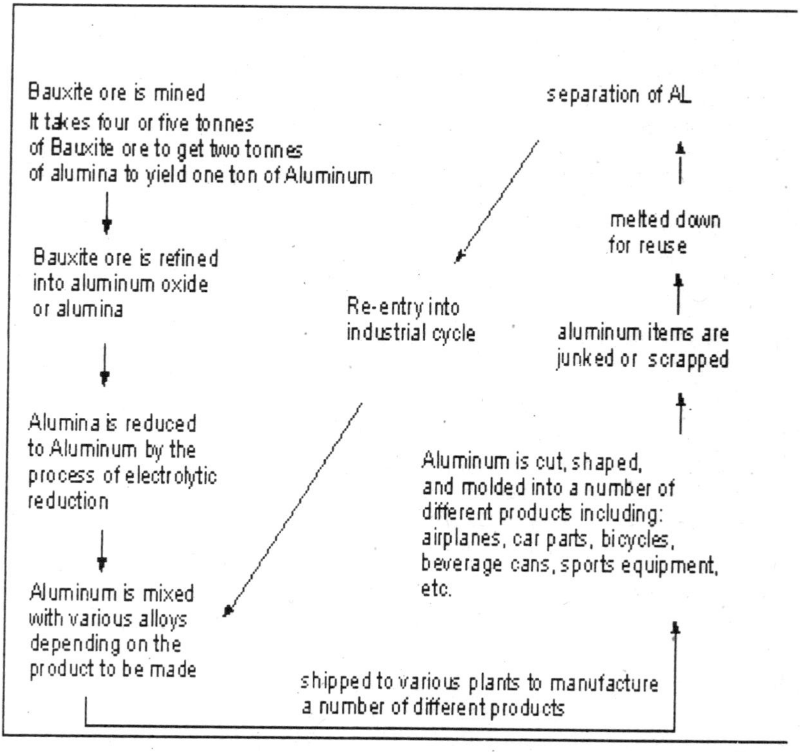

Table1. World Production of some Metals and mine waste in 1000 metric tonnes (KMT) (adapted from Ayres 1996)

Metal	*Weight of Ore KMT*	*Weight of Metal KMT*	*Mine & Mill* Waste
Aluminum	97,660	36,400	61,260
Copper	750,000	8,450	740,000
Iron	953,000	564,400	390,000
Uranium	1,900,000	36	1,900,000
Lead	101,000	3,380	98,000

Table 1 shows that each pound of aluminum extracted produces 1-7 pounds of waste at the mine, and almost all of the uranium ore is waste! The main intent of the table is to show the amounts of waste at the very first stage — mining and milling — in the production of a metal. Note that in all cases, we produce much more waste than metal. This is only the beginning. Each stage of production of a consumer good such as an automobile or a toaster, requires many more stages of manufacturing, each with its own wastes discarded to air, land, and water.

The Flows of Copper and Aluminum

The flow of copper is shown only from the mine to the point of input into an industrial process. The figure for Aluminum is drawn to represent the whole sequence of use, including recycling. These flow charts are the beginning stages of doing a life cycle analysis which we describe in the next section on Industrial Ecology. In the unit on Energy Systems, we have shown similar diagrams for the production of energy from different sources.

Copper

Over nine million metric tons (18,000 million lbs.) of copper is mined annually. The two main countries producing copper are Chile and the U.S., each producing about 2 million MT annually. Canada, the former USSR, Zaire, Zambia and Poland are the other countries. Copper is usually mined via open pits.

After a large amount of processing, each process with its other inputs and wastes, copper metal is delivered to the various industries that use it. Because copper is used in alloys such as brass and wires, it is difficult to extract and reuse the metal economically. Separation is one of the major obstacles in the attempt to recycle copper.

Copper is the oldest known metal to be extracted and used by humans. There are signs of copper use as early as 6000 B.C. toward the end of the Neolithic Age, considered by historians as the beginning of the Age of Metals. Alloying copper with tin or zinc to form bronze and brass seems to have happened by 3000 B.C. and other metallurgical heat-treatment processes like casting seems to have started by 1500 B.C. (Durant, Vol. I, p. 103).

Aluminum

Aluminum plays a very crucial role in modern society. It is the basis of aircraft and large mirrors used in telescopes! It is a versatile metal. The extraction and processing of aluminum from its ore - bauxite, Al_2O_3 — is environmentally destructive. Aluminum is not abundant in the earth's crust (14.3 atomic % or 8% by mass compared to 27.7% for Si (Silicon) and 5% for Fe (Iron), but it is in combination with oxygen. This strong combination requires a lot of energy to break!

The main producers of aluminum are Australia, Guinea, Jamaica and Brazil with much smaller amounts from Greece, France and Hungary. To separate aluminum (pure Al_2O_3) from the bauxite ore which also contains Fe_2O_3, SiO_2, TiO_2 and other minerals, a significant amount of electricity, water and chemicals such as limestone ($CaCO_3$) and Caustic Soda (NaOH) are needed. Separation of the metal Al from Al_2O_3 is done by means of an electrolytic process that consumes large amounts of electricity. Because of this, the companies producing aluminum had, in the early days, also developed hydroelectric power plants close to the mines. Thus the Tacurai Dam in Brazil sends one-third of its output electricity to aluminum smelters!

Because the extraction of Aluminum from the Alumina (Al_2O_3) ore requires an enormous amount of electrical energy, the aluminum industry initiated processes to recycle the used aluminum and was one of the first industries to do so. For many other materials, the sequence is not closed, that is, the used product is not processed to get the original material back.

Flows in the Case of a Consumer "Material" : Packaging

In the previous two cases, and in the material cycles earlier in the unit, we looked at the flows and cycles of materials that occur in nature. A large and diverse group of materials central to the consumer economy of today is all the material that could be categorized as packaging. The amounts and flows of packaging materials is an important aspect to consider because of their large effect on the environment.

Packaging including cartons, cans, and bottles used for consumer items are individually small but contribute to resource use and waste because of the sheer number of units we use. Packaging also has an often overlooked aspect; their centralized production and large distribution systems mean that we transport a large amount of consumer items over long distances, each with its own packaging. Thus transportation involves large expenditures of energy.

Packaging includes containers (made of glass, metals, or plastics), one-use containers (paper, plastic bags, cartons, and toothpaste tubes) and wrapping materials (paper, plastic, Aluminum foil). Recycling of packaging materials is very limited in the U.S. In 1993, we used about 310 kg (or 700 lbs.) per person of paper and paperboard for packaging, leading to a total of 80 million metric tons. Europe uses more plastics than paper because of the relative scarcity of land to grow wood.

As two examples of packaging materials, we now briefly describe the schemes of flow of paper and plastic. Paper and paperboard products are made mostly from wood pulp, and some paper from plant fibers such as cotton, linen and grasses.

Five thousand tons of paper are made per day in the U.S. to satisfy the 700 lbs per person per day demand. In 1996, 45% of post consumer waste paper was recycled, and the industry has set a 50% paper recovery goal. ("Industrial Environmental Performance Metrics," NAE Press, 1999.)

Increasingly, the industry in the U.S. is attempting to obtain wood grown on plantations or intensively managed forests rather than use wood from natural forests. Still, some of the tropical forests in South America and South East Asia are being harvested at alarming rates to provide wood for paper.

Paper pulping process uses high amounts of electricity and water — 2000 Kwh and 8000 gallons of water per ton of pulp produced. Large amounts of energy is also used when "pulp mats" are dried over hot rolls to become paper. The processing of paper, and especially the bleaching, consumes large amounts of chemicals. In particular, several million tons of chlorine are used which combines with the organic material in the process waste to produce over

1,000 types of organochlorine by-products. About 50 tons of organochlorines are released into water and air by an average size paper mill. Organochlorine byproducts are also concentrated in the sludge, 80% of which are buried in landfill. Environmental damage to plants, aquatic life and health risks to workers are well documented effects of paper mill effluents. The byproducts include dioxins which are both hazardous and bioaccumulative. Health and water pollution concerns have led the paper industry to look into chlorine free bleaching processes such as using ozone. A new family of bleaching compounds that decompose into water and oxygen has been developed by Professor Terry Collins of Carnegie Mellon University.

Pollution of the different media - air, water, and land - are discussed in the unit on the Atmospheric System, in the section on water cycle in this unit, and in the following section on solid and hazardous waste. Rethinking the use of materials using principles of ecology and "closing the circle" (a term used first by Barry Commoner) is the topic of the next section on Industrial Ecology.

Industrial Ecology

Industry is characterized as an economic activity in which materials are taken in, transformed and disbursed as products and dispensed of as waste. Over the last two centuries, technology has afforded us many ways of transforming materials so we have new, long-lived materials both as products and as waste.

The natural material cycles described in a previous section are part of our ecosystem. Over billions of years, materials, energy and life have all evolved to part of a natural ecology. In recent years, environmental problems and resource depletion have led to the rethinking of industrial uses of materials. The previous section gives some details of material use in industry. Whereas it has been customary to dispose of materials at the end of their use, industry especially in the U.S. and Europe, has begun to move toward a better accounting of material use, to exploring ways of using less energy and materials in production, and how to recycle materials and energy.

The scheme of industrial use of materials which tries to conserve and recycle materials is called Industrial Ecology. Industrial Ecology represents significant changes of mindset - or paradigm shift - from the previous centuries since the Industrial Revolution. Prior to this philosophy, we extracted and disposed of materials from nature as though nature provided us with an inexhaustible supply and as though nature could simply absorb increasingly large quantities of waste materials, some of which could not be degraded by natural processes.

The first move was to remove pollutants from the environment or to reduce pollutant emissions rather than to effect whole systemic changes. Resource depletion, air pollution, and the shortage of land and energy led to the first examination by industry of how to recover material from used items rather than keep extracting new material from earth.

A short review of Industry from the early economic vantage point and from the new ecological vantage point will give more background to the development of Industrial Ecology and decision making techniques that have evolved from the new paradigm of Industrial Ecology. Throughout the next sections, Industrial Ecology may be referred to as I.E.

Industry as an Ecological System

As the outlook changed from waste management as the environmental objective to pollution prevention, looking at the "ecology" of and industry in a group of industries began to be advanced. Graedel and Allenby, (1994) in the first textbook published on *Industrial Ecology*, writes:"Industrial Ecology is the means by which humanity can deliberately and rationally approach and maintain a desirable carrying capacity, given continued economic, cultural, and technological evolution. The concept requires that an industrial system be viewed not in isolation from its surrounding systems, but in context with them. It is a systems view in which one seeks to optimize the total materials cycle from virgin material, to finished material, to component, to obsolete product, and to ultimate disposal. Factors to be optimized include resources, energy, and capital." (Graedel and Allenby)The "deliberate", "internal" approach means that we need to understand the underlying science, technology and decision-making aspects so that each of these can be adjusted and the whole system optimized, not just for performance of the product to meet the needs of the user, but also to minimize the environmental damage caused by the life cycle of the product. The systems approach would help us make a transition from thinking of industry solely as an economic system to a more holistic framing of the industry as an ecology, compatible with the ecology of nature.In the unit on Ethical Systems, we discuss the gradual dawning of the philosophy of Industrial Ecology. (I.E.) The main focus of I.E. is to look at the whole system of production, use and disposal. The major focus of I.E., writes Ayres, is to:"identify opportunities for reducing wastes and pollution in the materials-intensive sections by exploiting opportunities for using the low-value byproducts (i.e. wastes) of certain processes as raw materials for others. Technical feasibility is the primary criterion for initial consideration." This view is an engineering perspective.A more global approach to Industrial Ecology (I.E.) would be to plan and carry out all industrial and related activities including consumer use, in a way as to minimize environmental harm, to think of ecology as a guiding principle for our economic-technological systems. Very briefly, the principles of nature's ecology are: cycling, a web of interaction and interdependence among parts; stability, and diffuse boundaries. These are discussed in more detail in the unit on Ecological Systems, scheme of the entire industrial - ecological system as represented by Bob Ayres and Kneese in the early 1970's. It is a beautiful representation of this system. It shows the two basic inputs: sunlight incorporated into natural materials through photosynthesis, and minerals from the earth. Starting with those two inputs, and adding air and water, we get quite a complex system of materials and energy in our industrial society. Note the large amounts of "residuals", or wastes.

This flow of materials and energy is called Industrial Metabolism, taking the analogy from biology. "Metabolism" of materials in industry uses energy, labor, capital and information or knowledge.Putting all these ideas together, we have a new framework for the ecology of industry. In the book Ecology of Industry [link], Deanna Richards and Robert Frosh introduces this idea with the following statement."...industry's ecology is defined by the metabolism of materials (the flow of materials through industrial systems, including their transformations during flow); the use of energy, labor and capital; and the application of information or knowledge. A characteristic of ecological systems is that they evolve. The evolution of industrial systems and their use (and storage) of resources are affected by the introduction of new technologies, decisions made in design, preferences of consumers, regulating dictates, and the like...Indeed "industrial ecology" has become jargon for describing systems of production and consumption networks that have a minimal impact on the environment as a primary objective and have an overarching objective of environmentally sustainable economic expansion."

Industry as an Economic System

Industrial development, especially in the 19th and 20th centuries assumed an inexhaustible supply of materials, energy, air, and water (the earth as an endless source). Constant growth in the production of goods was valued and waste production was not even considered a problem. Production measured by the GNP or Gross National Product, or the total value of goods and services produced during a given time period is taken as the indicator of progress, development, and a nation's well-being.

The societal and industrial attitude toward the use of the earth for human activities. This scheme shown in Figure IE1 below led to a loading of air, water and land by various materials, both toxic and nontoxic.

This sequence of production and use of consumer products in general consists of energy inputs and waste outputs at each stage, with limited energy recovery from waste in some industries such as paper, and some recycling of materials in industries such as aluminum. Because economic rather than environmental concerns were at the forefront, and gasoline and diesel for transportation was available cheaply, many production facilities were centralized to take advantage of economies of scale. Large factories produced large quantities of material and products, and a highway and rail system network distributed these to points of use. Thus each transportation stage in the Figure IE1 consisted of an energy expenditure.

Traditional economies have considered an industry, corporation, or business firm as a system, defined by their inputs and products. In this view, inputs are capital, materials (including energy) and labor - things paid for by the corporation. Outputs are products - goods and services the corporation produces. In this view, a view of what the corporation had to account for in its expenditures, waste and byproducts and especially pollution were considered externalities — "external" to the system since the corporation did not have to pay for it.

This myopic view led to behaviors such as discarding waste materials into air, land, and water without much thought. The U.S. Environmental Protection Agency, created in 1970 by the National Environmental Protection Act of 1969, forced industries to consider their waste streams as something they would have to pay for and add to their list of expenditures or inputs. This began to "internalized" some of the externalities. Pollution became a "cost" to the corporation because it would have to pay for remediation and disposal of waste and pollution.

This expanded the notion of the industrial system to include waste management at the site of production. It took a while for this to extend to "wastes" like the overburdening left at coal mines, land use, and other "costs" to the environment and risks to the health — of people and of nature — such as loss of biodiversity or incident of chronic disease. The burdening of land, air, and water, especially near large population centers such as cities or large industrial zones led to shortages of land and pollution of air and water. With some of the waste material and emissions being toxic, the nineteenth and twentieth centuries began seeing new "plagues" such as black lung, asbestosis, emphysema, various forms of cancer, hormone disruptions, nervous system effects and reproductive effects. It is a well known fact that lead paint has had detrimental effects on children, but it wasn't until the 1980's that exposure to lead in the air from gasoline was proven to caused IQ deficits in children. [Needleman] The high cost of caring for these new plagues and health related problems generated interest in Industrial Ecology by many unrelated industries. It is now starting to hit home that health care providers must pay the costs of injured or ill workers due to toxic exposure to pollution.

Resource depletion, air pollution, and the shortage of land and energy led to the first examination by industry of how to recover material from used items rather than continuing to extract new material from earth. The 1970's also saw "embedded energy analysis", a technique to see how much energy was used in our activities including product manufacture and use. A closer look at "life cycle of products" began in the mid-1980's. Figure IE2 shows a slightly altered scheme of production that began to emerge, with some attempts at material recovery.

Decision Making Techniques of Industrial Ecology

The basic premises of industrial ecology are resource conservation and pollution prevention, and achieving this on a system wide basis. Systems thinking means that we think of the product in terms of the materials extraction - refining - manufacturing - distribution - consumption - disposal scheme as we plan to produce it. We can then think of the different levels of the system at which industrial ecology could operate. Decision making techniques have been developed over the past decades that help to create a system wide industrial ecology in any community. There are four techniques that can be employed to create Industrial Ecology in any community.

Industry Lèvel Exchange: An industry that has a large list of input materials, or industries among whom there is a scope for exchange of materials can build a network to exchange materials.The automobile industry or a large producer of multiple products like 3M, IBM or General Electric can and do include their supplies in their plans on reusing and recycling materials. Large buyers of plastic like IBM, for example, get all their plastic from one supplier and buy such large quantities that they can set up a cycle of disassembling products and returning the plastic to the supplier. This type of "closing the loop" can also be of "economic" advantage to the firms or corporations, from a National academy of Engineering report illustrates a simplified process for automobile materials that show reuse and recycling of metallic and other parts. Automobile recycling has developed to be one of the most advanced systems, with only a fraction of the total materials going to landfills.

System of loosely coupled exchanges or a "community of exchange

"A firm may also look at its production system, and its input suppliers as well as other relevant organizations in its vicinity to see what kind of networks it can set up. A great example of this, cited often is the symbiotic relationship among a set of industries in Kalundborg, Denmark. A power plant, an oil refinery, a plasterboard manufacturing plant, a biotechnology plant, and a fish farm and all utilize material wastes from one another. Refined waste wastewater is used to cool the power plant. Excess gas from the refinery and sulfur obtained from the refinery are used to manufacture plaster board; biological sludge from the biotechnology plant is used by farmers; steam from the power plant is used by the biotechnology plant; waste heat from the power plant is used in fish farming and by the municipality for heating. One could conceivably design a community of process plants and other facilities to achieve similar symbiosis. Industrial Ecosystem.Reprinted from Measures of Environmental Performance and Ecosystem Condition. Copyright 1999 by the National Academy of Sciences. Courtesy of the National Academy Press, Washington, D.C. Reduce - Reuse - Recycle is a slogan that popularizes a waste management hierarchy of the "3R's" and is used widely for public awareness. This slogan was developed mainly for consumers as a way of thinking about their solid waste. But it could also be the cry for industry, especially in the U.S.The U.S. uses more resources than any other country in the world. Though we represent 5% of the world's population, we use 24% of the world's energy and 20% of material resources such as copper, tin, and lead. (ZPG Fact Sheet) The United States "consumed" more minerals between 1940 and 1976 than did all humanity up to 1940.The industrial ecology approach can be illustrated by a waste management system such as the sewage treatment system described above. Or, it can be applied to the larger scale design of a consumer product considering the pollution impacts across the entire life cycle of the product from raw materials to final disposal. This holistic "design for the environment" may not only minimize environmental impacts, but help improve the overall profitability of the product. Sewage Treatment SystemVarious public agencies have also attempted to set up material exchange programs. For example, in Sonoma County, California, the Public Education Conservation office in the county's Waste Management Agency facilitate exchanges through a quarterly newsletter as well as a website called SonoMax listing local materials avaukavke and wanted for individuals and businesses.

Environmental Accounting

The ultimate "challenge for corporations is to fully integrate environmental thinking into corporate decision making - to, in other words, translate their environmental concerns into the language of business", writes Ditz, Rapanakia and Banks in their book, Green Ledgers (anobib)[link] . As stated earlier, we have long considered industry in solely economic terms and all business ledgers have been set up to reflect costs, expenditures, and profits. The different sectors of the economy; some environmental costs are reflected in a firm's accounts, for example, the costs to clean the wastewater or gaseous effluents before they are discharged to the environment, or the costs of solid waste disposal. These costs have become part of the accounting or "internalized" since the enactment of the various environmental laws especially since 1970.Environmental costs in full, however, are more than what the firm remunerates. It is the cost of lost land and resources, loss of biodiversity, long term impacts of pollution, and health risks to the environment and people. Figure DM4 shows the representation in Ditz et al. of how the boundary of the private costs to industry are shifting. While some social costs of pollution have been internalized as cost of complying with regulation, other social costs, less easy to translate into monetary terms, remain outside the system boundary apparent to the corporation.The Shifting Boundary between Private and Social Costsadapted from Green Ledgers: Case Studies in Corporate Environmental Accounting by the World Resources Institute. May 1995.Several authors including *Carnegie Mellon*'s economist Lester Lave have suggested ways to bring environmental costs into a corporation's ledger by applying the idea of "full-cost accounting". In the accounting profession, full-cost accounting is the practice where the complete costs of the firm are incorporated into the pricing of the products. These authors suggest that this idea could be used to add the now external environmental costs into the decision making of the corporation. This means that the price of the product should reflect the entire private and social costs throughout the life cycle of the product, from raw material extraction to product disposal.What would happen if a firm did this? Their product would cost more. In the current way of thinking of the public, this might put the firm at a competitive disadvantage because only a few members of the public are willing to pay extra for a product whose manufacturer tries to compensate for the social and environmental costs due to the product's lifecycle. Full-cost accounting is, however, a method to explore as a way of including environmental costs in a company's ledger.

Life Cycle Analysis

Life Cycle Analysis (LCA) is a technique used to assess the environmental impacts of a product over its entire life cycle. Figure DM5a depicts the stages in a product life cycle. Figure DM5b outlines a flow chart representing the same stages. As discussed in the section on industrial uses of materials, these life "cycles" have not really been cycles but linear changes.

Stages in a Product Life Cycle

LCAs are used to evaluate the environmental impacts of products at every stage from material extraction to disposal of the used product. In general, a LCA consists of three components.

1) *Life Cycle Inventory:*

An account of the inputs and outputs at each stage. This would quantify the energy and raw materials that go into each stage, including transportation and outputs of products as well as all environmental releases.

2) *Life Cycle Impact Analysis:*

This would characterize the environmental loading and assess what ecological and human health impacts each loading would cause.

3) *Life Cycle Improvement Analysis:*

This analysis would systematically assess how the environmental loading and impacts could be reduced, without losing the quality of the product. The Society of Environmental Toxicology and Chemistry (SETAC) has provided an exhaustive description of the LCA methodology in their report, *A Technical Framework for Life Cycle Assessment.* Numerous authors including Robert Frosh and Deanna Richards, Robert Ayres, Robert Socolon, Brad

Allenby and Tom Graedel have written extensively on the topic. A LCA is hard to do because of all the data that are needed to do a complete job. The analysis can only be used for guidance rather than decision making because the available data might dominate the decision while the largest impacts might be from quantities for which the date are unavailable. All kinds of questions also arise such as where one starts the analysis, especially when comparing two alternatives. For example, when comparing glass bottles and plastic bottles, do we start the analysis with sand (raw material for glass) and oil platforms (as oil is the raw material for plastic), or assume that the glass and plastic are being manufactured anyway and to start with the glass and plastic as the "raw material"? Despite these problems, LCA provides a good framework for discussion of alternatives, and for detailed analysis when data are available.The sections below elaborate the details of a LCA.

1. *Life Cycle Inventory:* Figure DM6 from the SETAC report shows the scheme of inputs and outputs for a LCA. The stages of LCA inventory are:· Definition of system and system boundaries· List of raw materials, their sources, energy involved in extraction, wastes and effluents produced· Steps of processing the raw materials, stages involving combination of raw materials and manufacturing process· Possibilities for recycling materials during processing and manufacture· Accounting of energy and effluents from each of these steps ·Distribution and Transportation needed for the product to reach the consumer· Energy used and material waste and effluents produced during use and maintenance· Possibilities of reuse of whole product or parts· Possibilities of recycling of materials and the energy expenditure and effluent production in the recycling process.

The two brief examples below point our some of the

difficulties in doing the LCA to compare the environmental performance of two products. The first demonstration of the life cycle approach for a consumer product is the comparison of a paper cup and a Styrofoam cup described by Martin Hocking in his 1991 paper.(reference/bib) The use of Styrofoam for fast foods and hot beverages became a debate issue in the mid to late 1980s. The issue was that many people felt that Styrofoam would not degrade as easily as paper, and therefore was more of an environmental problem. Here, the primary focus was on the disposal stage of that particular consumer product. Hocking performed a life cycle analysis on the hot beverage cup and showed that the environmental issue is more complicated than just the disposal stage as follows:· Styrofoam relies on a non renewable resource (petroleum) for raw materials, paper use a renewable resource if the trees are from a sustainable tree farm· Similar high energy use for both products in the processing stage· Similar high water use for both in the processing stage· Paper was slightly heavier than styrofoam in terms of packaging and subsequent transportation· No differences in terms of use· In terms of disposal, paper has more mass than Styrofoam. Though paper is biodegradable, that biodegradability decreases significantly because of conditions in a landfill and because the paper has to be coated with wax for this use. In terms of incineration, Styrofoam has a greater energy potential. Final analysis is that it is hard to determine which is the better environmental choice since it is dependent of programs that are in place for the raw materials, use, and disposal. Another classic example is related to the energy source for automobiles i.e., electric versus gasoline engines. The 1990 amendments to the Clean Air Act required that several regions (such as Southern California, etc.) that were not meeting the tropospheric ozone standard, needed programs in place to ensure 2% of cars in 1998 were zero emission. The requirements increase over time. The trend has been towards the development of cars that are powered by electric batteries. These cars do not emit any emissions from the tailpipe and are declared zero emission vehicles. However, the "xero emission" is only because the focus of the regulatory strategy was on the use stage of the life cycle. The issue becomes more complicated when the entire life cycle is evaluated as several researchers have done. A simplified analysis is as follows: Raw materials for electric cars depend on the electricity source but is primarily coal. The electric cars also need a substantial battery that is currently made out of lead. Gasoline vehicles rely on petroleum. All of these raw materials are non renewable, though they may have different lifetimes. Lead emissions from lead processing plants cause air pollution with severe toxic effects (reference Hendrickson, Lave, and McMillian. Both power sources for the vehicles include high energy needs and water/air emissions during the refining of the raw materials, and the processing of the final product. Packaging systems differ drastically from transmission lines and batteries to tanker trucks, pipelines, and underground storage tanks. All have their associated environmental negatives. In terms of use, the electric car has zero emissions compared to the significant tailpipe emissions from current gasoline vehicles that are highly inefficient. However, there are emissions at the centralized electricity power plant that provides the electric vehicle with energy. In terms of disposal, there is the issue of battery disposal versus oil disposal. The better environmental product depends on several management choices concerning the energy source for the electricity, the battery material used, the location of the motor vehicles versus the centralized power plants, and so on. The issue is complicated.

2. *Life Cycle Impacts*. Following an inventory we could, in theory, assess what impacts the product will have in it's life cycle. Again this poses a number of problems as discussed in the two examples above.In assessing impacts, we need

to list and prioritize the impacts of concern - is it land loss, water pollution, global climate change, deforestation, human or ecological health hazards or all of these that make our list of concerns?For each type of emission or loading to the environment from the inventory analysis one could think of mapping the type of effects and the extent of the problem. A brief sketch of this is shown in Figure DM5. Note that this is just a representation to provide one way of depicting inventory and impact for a LCA.

A serious problem in calculating impacts is that the quantity of a material does not fully represent the human health or ecological impacts. Small amounts of dioxin are much more lethal than tons of a relatively benign substance like calcium oxide. The Green Design Initiative at Carnegie Mellon has addressed this by a clever scheme called the "CMU-ET" method.{missing part

Life Cycle Improvement Analysis: Green Design: Doing a LCA and it's use in Decision Making

Green Design maybe be defined as "Design that attempts to minimize environmental burdens without compromising functionality". The principle underlying green design is to assess life cycle impacts of the product during the design phase and add the minimization of environmental burdens as a design criteria to the usual design criteria such as performance, economy, reliability and safety. Aspects of green design include selecting materials that are more environmentally friendly (easy to recycle, little to no toxic byproducts), design for disassembly to facilitate reuse and recycling and designing for energy efficiency.Several modeling systems including the EPS or Environmental Priority Strategies System for products design initiated by the research institute of the Volvo Car Company in Sweden in 1990 have sought to construct databases to enable LCA calculations. There are numerous problems with getting exact numbers because of reasons such as incomplete accounting, material losses and proprietary nature of data. The Handbook of Industrial Energy Analysis by Ian Bokstead and G. F. Hancock of the Open University in England (John Wiley, 1979) and their database is probably the most extensive source. Many industries keep their own databases, and are sometimes willing to share information. Looking at a life cycle, one can think of generic strategies to adopt during design. In a World Resources Institute publication Ditz and Ranganathan (1997) developed four categories of measures that can be used to characterized a product:(1) Materials Use: quantities and types(2) Energy Consumption: quantities and types used or generated(3) Nonproduct output(4) Pollutant releasesCalculating these per unit of product is at the basis of every LCA. Design for the environment or Green Design then uses this data to evaluate and improve designs by several possible strategies: (1) Materials reduction or substitution and recycling (2) Energy reduction or substitution(3) Pollutant reduction and change in their natureExamples of this could be substitution of materials by those that would do the job with less toxic emissions in the production and use phases and designs that would use less energy in the use phase. Figure DM8 from the Richards and Frosch article cited before shows how the environmental considerations of lifecycle stages can lead to Green Design.insert DM8. Environmental considerations of lifecycle stagesThe student exercise on LCA [link] is an example of how despite the uncertainties a LCA could be used to compare alternative consumer products for making the "environmentally friendly decision on what products to buy. The exercise also points up the difficulties in such analysis.The Environmentally Responsible Product Assessment Matrick (Figure DM9) has been suggested by Laudise and Gaedel as a formulation for doing an evaluation. Each square of the Figure DM9 would have a checklist from which an impact analysis as in Figure DM7 could be done. Environmentally Responsible Product Assessment MatrickReprinted with permission from *The Ecology of Industry: Sectors and Linages*. Copyright 1998 by the National Academy of Sciences. Courtesty of the National Academy Press, Washington, D.C.Design for the environment (DFE) or Green Design as formulated by the electronics industry considered the following aspects in the design of electonic products such as computers:· design for disassembly or separability: how easy (and economical) are the components to separate for reuse or recovery of materials? design for recyclability: Is there potential for maximum recycling of component after use of the product? design for reusability: can components be reused in different product lines after recovering and refurbishing them? (Kovlak's reusable camera is an example of where this is practiced by the manufacturer) · design for remanufacture: can materials be recovered and recycled after use? this might include - setting up a system for consumers to send back used items. · design for disposability: can all materials and component be disposed of safely? Other aspects of DFC are discussed by Hutchinson, et al. at Green DesignFinally, we provide two examples to illustrate some of the ideas discussed above. Table DM1 is an inventory of the major raw materials, products, and product uses of the chemical industry as a whole. The graphs in Figure DM10 provide an example of the environmental profile of a computer workstation. These are both examples from the National Academy of Engineering Report, "Industrial Environmental Performance Metrics (NAE Press, 1999).

Raw MaterialsAirOilCoalWood Energy Sulfur Minerals Seawater Natural Gas Products Acids Nylon Alcohols Pigments/ dyes Benzene Polyester Caustic soda Polyethylene Esters Polyvinyl chloride Ethylene Solvents FibersSynthetic rubber Xylene Product End Users Adhesives Food ingredients Pharmaceuticals Automobiles

Fuel additives Piping Boats Household materials Preservatives Carpets Insulation Roofing Computers Packaging Safety glass Construction materials Paint and coatings Soaps and detergents Containers Paper Sports equipment Cosmetics Personal Care Textiles Fertilizers Pesticides Toys Tires. Major Raw Materials, Products and Product End Uses of the Chemical Industry Reprinted with permission from Industrial Environmental Performance Metrics Copyright 1998 by the National Academy of Sciences. Courtesty of the National Academy Press, Washington, D.C.

Environmental Profile of a Computer Workstation

Results of a life cycle study of the computer workstation are summarized in the graphs above. The computer workstations studies was assumed to contain one 1/6 inch thick silicon wafer (about 28 square inches), 220 integrated circuits (213 in plastic and 7 in ceramic packages), about 500 square inches (3.6 square feet) of single and multilayer printed wiring board, and a 20 inch monitor. The subcomponents included in the study were semiconductor devices (SD), semiconductor packaging (SP), printed wiring boards and computer assemblies (PWB/CA), and display units (Dis). The profiles of energy, material, and water use and waste reveal some aspects of the environmental impacts of an electronics product.

Conclusions

Frosch and Richards provides an overview of the progress being made by industry in environmental design and management. As discussed in the Unit on the Ethical System, the sustainability ethic is a goal towards which this figure optimistically points.As consumers, we have a responsibility to choose products that ensure a better environment and this implies an understanding of the complex system of natural and artificial materials that provide our needs, wants, and comforts. Industry's environmental design and management learning curveReprinted with permission from *Industrial Environmental Performance Metrics* Copyright 1998 by the National Academy of Sciences. Courtesty of the National Academy Press, Washington, D.C.

Carbon cycle

Diagram of the carbon cycle. The black numbers indicate how much carbon is stored in various reservoirs, in billions of tons ("GtC" stands for GigaTons of Carbon and figures are circa 2004). The purple numbers indicate how much carbon moves between reservoirs each year. The sediments, as defined in this diagram, do not include the ~70 million GtC of carbonate rock and kerogen.

The *carbon cycle* is the biogeochemical cycle by which carbon is exchanged between the biosphere, geosphere, hydrosphere, and atmosphere of the Earth.

The cycle is usually thought of as four major reservoirs of carbon interconnected by pathways of exchange. The reservoirs are the atmosphere, the terrestrial biosphere (which usually includes freshwater systems and non-living organic material, such as soil carbon), the oceans (which includes dissolved inorganic carbon and living and non-living marine biota), and the sediments (which includes fossil fuels). The annual movements of carbon, the carbon exchanges between reservoirs, occur because of various chemical, physical, geological, and biological processes. The ocean contains the largest active pool of carbon near the surface of the Earth, but the deep ocean part of this pool does not rapidly exchange with the atmosphere.

The *global carbon budget* is the balance of the exchanges (incomes and losses) of carbon between the carbon reservoirs or between one specific loop (e.g., atmosphere - biosphere) of the carbon cycle. An examination of the carbon budget of a pool or reservoir can provide information about whether the pool or reservoir is functioning as a source or sink for carbon dioxide.

In the atmosphere

Carbon exists in the Earth's atmosphere primarily as the gas carbon dioxide (CO_2). Although it is a very tiny percent of the atmosphere (approximately 0.04% on a molar basis, though rising), it plays an important role in supporting life. Other gases containing carbon in the atmosphere are methane and chlorofluorocarbons (the latter is entirely anthropogenic). The overall atmospheric concentration of these greenhouse gases has been increasing in recent decades, theoretically contributing to global warming.[1]

Carbon is taken from the atmosphere in several ways:

- When the sun is shining, plants perform photosynthesis to convert carbon dioxide into carbohydrates, releasing oxygen in the process. This process is most prolific in relatively new forests where tree growth is still rapid.
- At the surface of the oceans towards the poles, seawater becomes cooler and more carbonic acid is formed as CO_2 becomes more soluble. This is coupled to the ocean's thermohaline circulation which transports dense surface water into the ocean's interior (see the entry on the solubility pump).
- In upper ocean areas of high biological productivity, organisms convert reduced carbon to tissues, or carbonates to hard body parts such as shells and tests. These are, respectively, oxidized (soft-tissue pump) and redissolved (carbonate pump) at lower average levels of the ocean than those at which they formed, resulting in a downward flow of carbon (see entry on the biological pump).

- The weathering of silicate rock. Carbonic acid reacts with weathered rock to produce bicarbonate ions. The bicarbonate ions produced are carried to the ocean, where they are used to make marine carbonates. Unlike dissolved CO_2 in equilibrium or tissues which decay, weathering does not move the carbon into a reservoir from which it can readily return to the atmosphere.

Carbon can be released back into the atmosphere in many different ways:

- Through the respiration performed by plants and animals. This is an exothermic reaction and it involves the breaking down of glucose (or other organic molecules) into carbon dioxide and water.
- Through the decay of animal and plant matter. Fungi and bacteria break down the carbon compounds in dead animals and plants and convert the carbon to carbon dioxide if oxygen is present, or methane if not.
- Through combustion of organic material which oxidizes the carbon it contains, producing carbon dioxide (and other things, like water vapor). Burning fossil fuels such as coal, petroleum products, and natural gas releases carbon that has been stored in the geosphere for millions of years. Burning agrofuels also releases carbon dioxide.
- Production of cement. Carbon dioxide is released when limestone (calcium carbonate) is heated to produce lime (calcium oxide), a component of cement.
- At the surface of the oceans where the water becomes warmer, dissolved carbon dioxide is released back into the atmosphere.

- Volcanic eruptions and metamorphism release gases into the atmosphere. Volcanic gases are primarily water vapor, carbon dioxide and sulfur dioxide. The carbon dioxide released is roughly equal to the amount removed by silicate weathering; so the two processes, which are the chemical reverse of each other, sum to roughly zero, and do not affect the level of atmospheric carbon dioxide on time scales of less than about 100,000 yr.

In the biosphere

Around 1,900 gigatons of carbon are present in the biosphere. Carbon is an essential part of life on Earth. It plays an important role in the structure, biochemistry, and nutrition of all living cells.

- Autotrophs are organisms that produce their own organic compounds using carbon dioxide from the air or water in which they live. To do this they require an external source of energy. Almost all autotrophs use solar radiation to provide this, and their production process is called photosynthesis. A small number of autotrophs exploit chemical energy sources in a process called chemosynthesis. The most important autotrophs for the carbon cycle are trees in forests on land and phytoplankton in the Earth's oceans. Photosynthesis follows the reaction $6CO_2 + 6H_2O$ ' $C_6H_{12}O_6 + 6O_2$
- Carbon is transferred within the biosphere as heterotrophs feed on other organisms or their parts (e.g., fruits). This includes the uptake of dead organic material (detritus) by fungi and bacteria for fermentation or decay.
- Most carbon leaves the biosphere through respiration. When oxygen is present, aerobic respiration occurs,

which releases carbon dioxide into the surrounding air or water, following the reaction $C_6H_{12}O_6 + 6O_2$ '$6CO_2 + 6H_2O$. Otherwise, anaerobic respiration occurs and releases methane into the surrounding environment, which eventually makes its way into the atmosphere or hydrosphere (e.g., as marsh gas or flatulence).

- Burning of biomass (e.g. forest fires, wood used for heating, anything else organic) can also transfer substantial amounts of carbon to the atmosphere
- Carbon may also be circulated within the biosphere when dead organic matter (such as peat) becomes incorporated in the geosphere. Animal shells of calcium carbonate, in particular, may eventually become limestone through the process of sedimentation.
- Much remains to be learned about the cycling of carbon in the deep ocean. For example, a recent discovery is that larvacean mucus houses (commonly known as "sinkers") are created in such large numbers that they can deliver as much carbon to the deep ocean as has been previously detected by sediment traps.[2] Because of their size and composition, these houses are rarely collected in such traps, so most biogeochemical analyses have erroneously ignored them.

Carbon storage in the biosphere is influenced by a number of processes on different time-scales: while net primary productivity follows a diurnal and seasonal cycle, carbon can be stored up to several hundreds of years in trees and up to thousands of years in soils. Changes in those long term carbon pools (e.g. through de- or afforestation or through temperature-related changes in soil respiration) may thus affect global climate change.

In the ocean

"Present day" (1990s) sea surface dissolved inorganic carbon concentration (from the GLODAP climatology)

The oceans contain around 36,000 gigatonnes of carbon, mostly in the form of bicarbonate ion. Inorganic carbon, that is carbon compounds with no carbon-carbon or carbon-hydrogen bonds, is important in its reactions within water. This carbon exchange becomes important in controlling pH in the ocean and can also vary as a source or sink for carbon. Carbon is readily exchanged between the atmosphere and ocean. In regions of oceanic upwelling, carbon is released to the atmosphere. Conversely, regions of downwelling transfer carbon (CO_2) from the atmosphere to the ocean. When CO_2 enters the ocean, carbonic acid is formed:

$$CO_2 + H_2O \rightleftharpoons H_2CO_3$$

This reaction has a forward and reverse rate, that is it achieves a chemical equilibrium. Another reaction important in controlling oceanic pH levels is the release of hydrogen ions and bicarbonate. This reaction controls large changes in pH:

$$H_2CO_3 \rightleftharpoons H^+ + HCO_3^-$$

In the oceans, bicarbonate can combine with calcium to form limestone (calcium carbonate, $CaCO_3$, with silica), which precipitates to the ocean floor. Limestone is the largest reservoir of carbon in the carbon cycle. The calcium comes from the weathering of calcium-silicate rocks, which causes the silicon in the rocks to combine with oxygen to form sand or quartz (silicon dioxide), leaving calcium ions available to form limestone[3].

REFERENCES

1. The Intergovernmental Panel on Climate Change (IPCC) which represents wide consensus of

international scientific opinion.

2. Monterey Bay Aquarium Research Institute (MBARI) (2005-06-09). ""Sinkers" provide missing piece in deep-sea puzzle". Press release. Retrieved on 2007-10-07.

3. Notes, Lecture. The Carbon Cycle. Department of Atmospheric Sciences. University of Washington. Retrieved on 2007-12-21.

9

Earth's Atmosphere

"Air" redirects here. For other uses, see Air (disambiguation). Atmospheric gases scatter blue light more than other wavelengths, giving the Earth a blue halo when seen from space.

Nitrogen	78.0842%
Oxygen	20.9463%
Argon	0.93422%
Carbon dioxide	0.03811%
Water vapor about	1%
Other	0.002%

Layers of the atmosphere (not to scale)The Earth's atmosphere is a layer of gases surrounding the planet Earth and retained by the Earth's gravity. It contains roughly (by molar content/volume) 78.08% nitrogen, 20.95% oxygen, 0.93% argon, 0.038% carbon dioxide, trace amounts of other gases, and a variable amount (average around 1%) of water vapor. This mixture of gases is commonly known as air. The atmosphere protects life on Earth by absorbing ultraviolet solar radiation and reducing temperature extremes between day and night.

There is no definite boundary between the atmosphere

and outer space. It slowly becomes thinner and fades into space. Three quarters of the atmosphere's mass is within 11 km of the planetary surface. In the United States, people who travel above an altitude of 80.5 km (50 statute miles) are designated astronauts. An altitude of 120 km (~75 miles or 400,000 ft) marks the boundary where atmospheric effects become noticeable during re-entry. The Kármán line, at 100 km (62 miles or 328,000 ft), is also frequently regarded as the boundary between atmosphere and outer space.

Temperature and layers

The temperature of the Earth's atmosphere varies with altitude; the mathematical relationship between temperature and altitude varies among five different atmospheric layers (ordered highest to lowest, the ionosphere is part of the thermosphere):

Exosphere: from 500 – 1000 km (300 – 600 mi) up to 10,000 km (6,000 mi), free-moving particles that may migrate into and out of the magnetosphere or the solar wind. It plays an important part in atmospheric electricity and forms the inner edge of the magnetosphere. It has practical importance because, among other functions, it influences radio propagation to distant places on the Earth. It is located in the thermosphere and is responsible for auroras.

Thermosphere: from 80 – 85 km (265,000 – 285,000 ft) to 640+ km (400+ mi), temperature increasing with height.

Mesosphere: From the Greek word meaning middle. The mesosphere extends from about 50 km (160,000 ft) to the range of 80 to 85 km (265,000 – 285,000 ft), temperature decreasing with height. This is also where most meteors burn up when entering the atmosphere.

Stratosphere: From the Latin word "stratus" meaning

a spreading out. The stratosphere extends from the troposphere's 7 to 17 km (23,000 – 60,000 ft) range to about 50 km (160,000 ft). Temperature increases with height. The stratosphere contains the ozone layer, the part of the Earth's atmosphere which contains relatively high concentrations of ozone. "Relatively high" means a few parts per million—much higher than the concentrations in the lower atmosphere but still small compared to the main components of the atmosphere. It is mainly located in the lower portion of the stratosphere from approximately 15 to 35 km (50,000 – 115,000 ft) above Earth's surface, though the thickness varies seasonally and geographically.

Troposphere: From the Greek word meaning to turn or change. The troposphere is the lowest layer of the atmosphere; it begins at the surface and extends to between 7 km (23,000 ft) at the poles and 17 km (60,000 ft) at the equator, with some variation due to weather factors. The troposphere has a great deal of vertical mixing due to solar heating at the surface. This heating warms air masses, which makes them less dense so they rise. When an air mass rises the pressure upon it decreases so it expands, doing work against the opposing pressure of the surrounding air. To do work is to expend energy, so the temperature of the air mass decreases. As the temperature decreases, water vapor in the air mass may condense or solidify, releasing latent heat that further uplifts the air mass. This process determines the maximum rate of decline of temperature with height, called the adiabatic lapse rate. It contains roughly 80% of the total mass of the atmosphere. 50% of the total mass of the atmosphere is located in the lower 5 km of the troposphere.

The average temperature of the atmosphere at the surface of Earth is 15 °C (59 °F).[1]

Pressure and thickness

Barometric Formula: (used for airplane flight) barometric formula

The average atmospheric pressure, at sea level, is about 101.3 kilopascals (about 14.7 psi); total atmospheric mass is 5.1361×1018 kg [2].

Atmospheric pressure is a direct result of the total weight of the air above the point at which the pressure is measured. This means that air pressure varies with location and time, because the amount (and weight) of air above the earth varies with location and time.

Atmospheric pressure decreases with height, dropping by 50% at an altitude of about 5.6 km (18,000 ft). Equivalently, about 50% of the total atmospheric mass is within the lowest 5.6 km. This pressure drop is approximately exponential, so that pressure decreases by approximately half every 5.6 km. However, because of changes in temperature throughout the atmospheric column, as well as the fact that the force of gravity begins to decrease at great altitudes, a single equation does not model atmospheric pressure through all altitudes (it is modeled in 7 exponentially decreasing layers, in the equations given above).

Even in the exosphere, the atmosphere is still present (as can be seen for example by the effects of atmospheric drag on satellites).

The equations of pressure by altitude in the above references can be used directly to estimate atmospheric thickness. However, the following published data are given for reference:- [3]

50% of the atmosphere by mass is below an altitude of 5.6 km.

90% of the atmosphere by mass is below an altitude

of 16 km. The common altitude of commercial airliners is about 10 km.

99.99997% of the atmosphere by mass is below 100 km. The highest X-15 plane flight in 1963 reached an altitude of 354,300 ft (108,000 m).

Therefore, most of the atmosphere (99.9997%) is below 100 km, although in the rarefied region above this there are auroras and other atmospheric effects.

Composition

Composition of Earth's atmosphere as at 1987.Dec. The lower pie represents the least common gases that compose 0.038% of the atmosphere. Values normalized for illustration.

Mean atmospheric water vaporComposition of dry atmosphere, by volume[4] ppmv: parts per million by volume

Gas Volume

Nitrogen (N_2) 780,840 ppmv (78.084%)

Oxygen (O_2) 209,460 ppmv (20.946%)

Argon (Ar) 9,340 ppmv (0.9340%)

Carbon dioxide (CO_2) 383 ppmv (0.0383%)

Neon (Ne) 18.18 ppmv (0.001818%)

Helium (He) 5.24 ppmv (0.000524%)

Methane (CH_4) 1.745 ppmv (0.0001745%)

Krypton (Kr) 1.14 ppmv (0.000114%)

Hydrogen (H_2) 0.55 ppmv (0.000055%)

Not included in above dry atmosphere:

Water vapor (H_2O) 0.25% over full atmosphere, typically

1% to 4% near surface

Minor components of air not listed above include[citation needed] Gas Volume

nitrous oxide 0.3 ppmv (0.00005%)
xenon 0.09 ppmv (9x10-6%)
ozone 0.0 to 0.07 ppmv (0%-7x10-6%)
nitrogen dioxide 0.02 ppmv (2x10-6%)
iodine 0.01 ppmv (1x10-6%)
carbon monoxide trace
ammonia trace

The mean molar mass of air is 28.97 g/mol. Note that the composition figures above are by volume-fraction (V%), which for ideal gases is equal to mole-fraction (that is, fraction of total molecules). By contrast, mass-fraction abundances of gases, particularly for gases with significantly different molecular (molar) mass from that of air will differ from those by volume. For example, in air, helium is 5.2 ppm by volume-fraction and mole-fraction, but only about (4/29) × 5.2 ppm = 0.72 ppm by mass-fraction.

Heterosphere

Below the turbopause at an altitude of about 100 km (not far from the mesopause), the Earth's atmosphere has a more-or-less uniform composition (apart from water vapor) as described above; this constitutes the homosphere.[5] However, above about 100 km, the Earth's atmosphere begins to have a composition which varies with altitude. This is essentially because, in the absence of mixing, the density of a gas falls off exponentially with increasing altitude, but at a rate which depends on the molar mass. Thus higher mass constituents, such as oxygen and nitrogen, fall off more quickly than lighter constituents such as helium, molecular hydrogen, and atomic hydrogen. Thus there is a layer, called the heterosphere, in which the earth's atmosphere has varying composition. As the altitude increases, the atmosphere is dominated successively by helium, molecular hydrogen, and atomic hydrogen. The precise altitude of the heterosphere and the layers it contains varies significantly with temperature. After loss of the hydrogen, helium and other hydrogen-containing gases from early Earth due to the Sun's radiation, primitive Earth was devoid of an atmosphere. The first atmosphere was formed by outgassing of gases trapped in the interior of the early Earth, which still goes on today in volcanoes. [6]

Density and mass

Earth's atmosphere from space.

Temperature and mass density against altitude from the NRLMSISE-00 standard atmosphere model.Main article: Density of air

The density of air at sea level is about 1.2 kg/m^3(1.2 g/L). Natural variations of the barometric pressure occur at any one altitude as a consequence of weather. This variation is relatively small for inhabited altitudes but much more pronounced in the outer atmosphere and space due to variable solar radiation.

The atmospheric density decreases as the altitude increases. This variation can be approximately modeled using the barometric formula. More sophisticated models are used by meteorologists and space agencies to predict weather and orbital decay of satellites.

The average mass of the atmosphere is about 5 quadrillion metric tons or 1/1,200,000 the mass of Earth. According to the National Center for Atmospheric Research, "The total mean mass of the atmosphere is 5.1480×1018 kg with an annual range due to water vapor of 1.2 or 1.5×1015 kg depending on whether surface pressure or water vapor data are used; somewhat smaller than the previous estimate. The mean mass of water vapor is estimated as 1.27×1016 kg and the dry air mass as 5.1352 ±0.0003×1018 kg."

Evolution on Earth

The history of the Earth's atmosphere prior to one billion years ago is poorly understood and an active area of scientific research. The following discussion presents a plausible scenario.

The modern atmosphere is sometimes referred to as Earth's "third atmosphere", in order to distinguish the current chemical composition from two notably different previous compositions. The original atmosphere was primarily helium and hydrogen. Heat from the still-molten crust, and the sun, plus a probably enhanced solar wind, dissipated this atmosphere.

About 4.4 billion years ago, the surface had cooled enough to form a crust, still heavily populated with volcanoes which released steam, carbon dioxide, and ammonia. This led to the early "second atmosphere", which was primarily carbon dioxide and water vapor, with some nitrogen but virtually no oxygen. This second atmosphere had approximately 100 times as much gas as the current atmosphere, but as it cooled much of the carbon dioxide was dissolved in the seas and precipitated out as carbonates. The later "second atmosphere" contained largely nitrogen and carbon dioxide. However, simulations run at the University of Waterloo and University of Colorado in 2005 suggest that it may have had up to 40% hydrogen.[7] It is generally believed that the greenhouse effect, caused by high levels of carbon dioxide and methane, kept the Earth from freezing.

One of the earliest types of bacteria was the

cyanobacteria. Fossil evidence indicates that bacteria shaped like these existed approximately 3.3 billion years ago and were the first oxygen-producing evolving phototropic organisms. They were responsible for the initial conversion of the earth's atmosphere from an anoxic state to an oxic state (that is, from a state without oxygen to a state with oxygen) during the period 2.7 to 2.2 billion years ago. Being the first to carry out oxygenic photosynthesis, they were able to produce oxygen while sequestering carbon dioxide in organic molecules, playing a major role in oxygenating the atmosphere.

Photosynthesising plants would later evolve and continue releasing oxygen and sequestering carbon dioxide. Over time, excess carbon became locked in fossil fuels, sedimentary rocks (notably limestone), and animal shells. As oxygen was released, it reacted with ammonia to release nitrogen; in addition, bacteria would also convert ammonia into nitrogen. But most of the nitrogen currently present in the atmosphere results from sunlight-powered photolysis of ammonia released steadily over the aeons from volcanoes.

As more plants appeared, the levels of oxygen increased significantly, while carbon dioxide levels dropped. At first the oxygen combined with various elements (such as iron), but eventually oxygen accumulated in the atmosphere, resulting in mass extinctions and further evolution. With the appearance of an ozone layer (ozone is an allotrope of oxygen) lifeforms were better protected from ultraviolet radiation. This oxygen-nitrogen atmosphere is the "third atmosphere". 200 – 250 million years ago, up to 35% of the atmosphere was oxygen (as found in bubbles of ancient atmosphere were found in an amber).

This modern atmosphere has a composition which is enforced by oceanic blue-green algae as well as geological processes. O2 does not remain naturally free in an atmosphere, but tends to be consumed (by inorganic chemical reactions, and by animals, bacteria, and even land plants at night), and CO2 tends to be produced by respiration and decomposition and oxidation of organic matter. Oxygen would vanish within a few million years due to chemical reactions and CO2 dissolves easily in water and would be gone in millennia if not replaced. Both are maintained by biological productivity and geological forces seemingly working hand-in-hand to maintain reasonably steady levels over millions of years.

Air pollution

Before desulfurization filters were installed, the emissions from this power plant in New Mexico contained excessive amounts of sulfur dioxide.

Air pollution is a chemical, physical (e.g. particulate matter) or biological agent that modifies the natural characteristics of the atmosphere in an unwanted way. Stratospheric ozone depletion due to air pollution (chiefly from chlorofluorocarbons) has long been recognized as a threat to human health as well as to the earth's ecosystems.

Worldwide air pollution is responsible for large numbers of deaths and cases of respiratory disease. Enforced air quality standards, like the Clean Air Act in the United States, have reduced the presence of some pollutants. While major stationary sources are often identified with air pollution, the greatest source of emissions is actually mobile sources, principally the automobile. Gases such as carbon dioxide, methane, and fluorocarbons contribute to global warming, and these gases, or excess amounts of some emitted from fossil fuel burning, have recently been identified by the United States and many other countries (see Kyoto accord), as pollutants.[citation needed]

REFERENCES

Earth's Radiation Balance and Oceanic Heat Fluxes.

Global atmospheric mass, surface pressure, and water vapor variations

Lutgens, Frederick K. and Edward J. Tarbuck (1995) The Atmosphere, Prentice Hall, 6th ed., pp14-17, ISBN 0-13-350612-6

Source for figures: Carbon dioxide, NASA Earth Fact Sheet, (updated 2007.01). Methane, IPCC TAR table 6.1, (updated to 1998). The NASA total was 17 ppmv over 100%, and CO2 was increased here by 15 ppmv. To normalize, N2 should be reduced by about 25 ppmv and O2 by about 7 ppmv.

Vercheval, J. The thermosphere: a part of the heterosphere. (offline, see Internet Archive copy)

"Early Earth atmosphere favorable to life: study", University of Waterloo, April 7, 2005. Retrieved on 2007-07-30.

Bibliography

Agarwal, Anil, Sunita Narain and Anju Sharma (ed.), *Green Politics: Global Environmental Negotations,*Vol.II. (New Delhi: CSE, 2001)

Anil Agarwal, Sunita Narain & Anju Sharma (ed), *Global Environmental Negotiations,* Vol. I, (New Delhi: CSE, 1999).

'A Non Aggression pact with Nature,' *The Hindu,* June 3, 1990.

Bhagwati, J., *India in Transition-Freeing the Economy,* (Oxford: Clarendon Press 1993).

Brenton,Tony, *The Greening of Machiavelli: The Evolution of International EnvironmentalPolics,* (London: Earth Scan, 1994).

Christina Lamb, "US Presence at Earth Summit Vital Collar", *The Financial Times,* (London), March 27, 1992.

'Conservation of Natural Wealth Most Essential, *The Hindu* November 4, 1989.

Conference on Environment and Development', December 22, 1989.

Earth's Radiation Balance and Oceanic Heat Fluxes.

"Early Earth atmosphere favorable to life: study", University of Waterloo, April 7, 2005.

Gareth Porter and Janet Brown (eds), *Global Environmental Politics,* (Oxford: Westview Press, 1991).

Global atmospheric mass, surface pressure, and water vapor variations

Joint Comminque of the SAARC Minister of Environment New Delhi, April 9, 1992 and Kuala Lumpur Declaration on Environment and Development, Kuala Lumpur, April 29, 1992.

'Joint Statement of 9-77 at the Fourth Session of the INC for a FCCC, Geneva, December 9-20 1991.

'Looming Menace to the ozone layers', *The Hindu,* December 31, 1987.

Lutgens, Frederick K. and Edward J. Tarbuck (1995) The Atmosphere, Prentice Hall, 6th ed., pp14-17, ISBN 0-13-350612-6

Ministry of Environment and Forests. Annual Report 1984-85, New Delhi, 1985, p. I.

MOEF, *Second Meeting of the EAC on Global Environmental Issue,* New Delhi, May 24, 1990.

Mitra, A.P., (ed.), *Greenhouse Gas Emissions in India: A Preliminary Report,* New Delhi: CSIR, 1991.

M.G. Rajan, *Global Environmental Polices.' India and the North -South Politics of Global Environmental Issues,* (Delhi: Oxford University Press, 1997).

MOEF, 'Greenhouse Effect and Climate Change-Issues for the Developing Countries', New Delhi, April S, 1990.

MOEF, *Chairman's Summary Conference of Select Developing Countries on Global Environmental Issues,* New Delhi, April 23-25, 1990.

Monterey Bay Aquarium Research Institute (MBARI) (2005-06-09). ""Sinkers" provide missing piece in deep-sea puzzle". Press release. Retrieved on 2007-10-07.

Mohamed T.EL-Ashry, 'Statement to the Fifth INC-FCCC', (New York), February 22, 1992.

National Committee on Environmental Planning and Coordination, Inaguaral Function (proceedings published by the *Department of Science & Technology*, New Delhi, 12 April 1972.

Notes, Lecture. The Carbon Cycle. Department of Atmospheric Sciences. University of Washington. Retrieved on 2007-12-21.

O.P. Dwivedi, *India's Environmental Policies, Programmes and Stewardship,* (New York: St Martin's Press, 1997).

O.P. Dwivedi, "India: Pollution Control Policy and Programmes", *International Review of Administrative Sciences,* Vol., 43, no. 2, 1977.

Rahmatullah Khan, "Legal and Institutional Issues Arising out of the Proposed Framework Convention on Climate Change", in Yatendra Josho et.al (eds), *Global Climate Change: Science Impacts and Responses,* (New Delhi: TERI, 1992).

Policy Statement, Meeting of OECD Minister on Environment and Development, OECD, Paris, December 2-3, para II.

Session, Geneva, December 9-20, 1991, *AlAC.* 237/15, January 19, 1-992.

Source for figures: Carbon dioxide, NASA Earth Fact Sheet, (updated 2007.01). Methane, IPCC TAR table 6.1, (updated to 1998). The NASA total was 17 ppmv over 100%, and CO2 was increased here by 15 ppmv. To normalize, N2 should be reduced by about 25 ppmv and O2 by about 7 ppmv.

The process of preparing citizens' reports was pioneered in the region by the Centre for Science and Environment (CSE), in New Delhi, India. The State of India's Environment -The first Citizens Report was published in 1982. Since then CSE has been publishing citizens report on a regular basis.

The Intergovernmental Panel on Climate Change (IPCC) which represents wide consensus of international scientific opinion.

TERI, Report on Global Warming Anti Associated Impacts, (New Delhi: TERI), June 1990.

Tolba, Mostafa K., (ed.,) Evolving EnvironemntalPerceptions: From Stockholm to Nairobi,(London: Butterworth's, 1989).

UN General Assembly, "Protection of Global Climate for Present and Future Generations", Resolution *45/212,* December 21, 1990.

UNGA, Report of INC for FCCC on Climate Change on the work of its First Session, *A/AC,237/6,* March 8, 1991.

UNGA, Report of the INC for a FCCC on the work of its Fourth

UNGA, Resolution 2398 (XXIII)

UN, General Assembly, Resolution 44/228, United Nations

UNEP/WMO, Adhoc Working Group of Government Representatives to Prepare for Negotiations on a Framework Convention on Climate Change UNEP/

WMO Prep/FCCC/LI/ REPORT, Geneva, September 24-26, 1990.

UN Document, *AlAC. 2371* MISC. 20, February 28,1992.

UNEP, Program Planning Country Report - India, 1968, MIMEO, Nairobi, UNE, p. 2.

UN, General Assembly Report of the INC for a FCCC on the work of its Second Session, Held at Geneva from 19-28 June, 1991, *AlAC.* 237-9, August 19, 1991.

U GA, Report of the INC for a FCCC on the Work of its Third Session, Nairobi, September 9-20,1991, *A/AC* 237/12, October 25, 1991.

UNEP, Annual Report of the Executive Director 1991, Nairobi, UNEP, 1992.

Vercheval, J. The thermosphere: a part of the heterosphere. (offline, see Internet Archive copy)

White, Rodney R.,North South, and the Environmental Crisis, (London: University of Toronto Press,1993).

WRl: World Resources Institute, *World Resources -1990-91* (Oxford: Oxford University Press,1990).

WMO, Report of the meeting of Government Representative: Draft Ministerial Declaration for the Second World Climate Conference, WMO, Geneva, September 27-29, 1990.

Index

***